Tools of Radio Astronomy

Tools of Radio Astronomy

Editor

Izaan Ahmed

Tools of Radio Astronomy

Edited by **Izaan Ahmed**

Printed in 2017

ISBN: 978-1-68117-177-7
Library of Congress Control Number: 2015954089

© 2016 by
SCITUS Academics LLC,
616, Corporate Way, Suite 2, 4766,
Valley Cottage, NY 10989

www.scitusacademics.com

Preface

When you gaze up at the night sky you see light given off by stars. That light has travelled across space for dozens, hundreds or thousands of years before entering your eye. When astronomers use large telescopes to probe the Universe, the faint light they gather may have come from objects millions or billions of light years away. In effect, we see objects as they were in the past as it takes that light time to travel across space. Astronomy, perhaps the oldest of Sciences, is the study of celestial objects including the planets, stars, and galaxies - even the Universe as a whole. What then is radio astronomy?

Radio astronomy, a subfield of astronomy, studies celestial objects at radio frequencies. The initial detection of radio waves from an astronomical object was made in the 1930s, when Karl Jansky observed radiation coming from the Milky Way. Subsequent observations have identified a number of different sources of radio emission. These include stars and galaxies, as well as entirely new classes of objects, such as radio galaxies, quasars, pulsars, and masers. Radio astronomy is conducted using large radio antennas referred to as radio telescopes, that are either used singularly, or with multiple linked telescopes utilizing the techniques of radio interferometry and aperture synthesis.

This book gives a complete introduction to the instrumentation and techniques needed for radio-astronomical research. After a thorough survey of electromagnetic wave propagation, antenna theory and the design of receivers are dealt with. Radiation mechanisms relevant to radio astronomy are the also subject of the in this book.

Table of Contents

Table of Contents

Table of Contents

Table of Contents

Chapter 1

Radio Signatures of Sunspot NOAA 12192

MinttuUunila, JuhaKallunki
Metsähovi Radio Observatory, Aalto University, Kylmälä, Finland

ABSTRACT

In this paper, we present an overview of radio signatures of sunspot NOAA 12192 measured with various instruments with frequencies of 37 GHz, 11.2 GHz and 200 - 400 MHz at Aalto University Metsähovi Radio Observatory (MRO). The data were observed during October 20 - 29, 2014. In total, 12 solar radio bursts at 11.2 GHz and 8 at 200 - 400 MHz, with varying intensities and properties, were observed. Radio brightening was captured in several solar radio maps. NOAA 12192 is the largest observed sunspot during solar cycle 24. We show that this exceptional radio brightening belongs to the strongest category including less than 5% of radio brightenings ever measured at MRO.

INTRODUCTION

Sunspot NOAA 12192, Figure 1, was both active and visible during October 20 - 29, 2014. It is the largest sunspot, with a diameter > 100,000 km, measured in solar cycle 24 regarding the size of the umbra [1] , [2] , and also the largest one observed in 24 years [3] producing tens of solar events [4] . The high activity and unstability of NOAA 12192 shows that the sunspot has a highly complex magnetic structure. The sunspot, on the right side in Figure 1, shows also various flux structures measured with SDO/HMI magnetogram. The black areas have magnetic fields with negative polarity, and white areas have positive polarity. In typical sunspots, like 2194 in Figure 1on the right side, there is only one negative and one positive polarity area, and they are clearly separated from each other. In the case of NOAA 12192, marked as 2192 in Figure 1, there are several polarity areas. The negative magnetic field in the Figure 1 is −2.0 kG, and the positive magnetic field is more than 1.7 kG.

Large sunspots like NOAA 12192 are not usually observed, because they tend to lose their stability as they grow and change into more complex magnetic structure. They usually disintegrate into smaller sunspots according

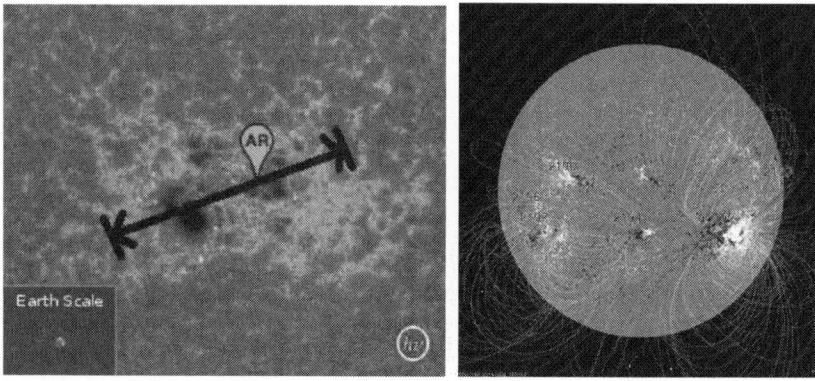

Figure 1: NOAA 12192 as seen by SDO AIA 1600 Å. Figure on the left is courtesy of NASA, ESA and JAXA. The marked area is > 100 000 km. Figure on the right with photosgraphic magnetic field map is measured by SDO/HMI magnetogram (at 6173 Å) on October 26, 2014 at 22:01:59 UT, courtesy of NASA/SDO and the AIA, EVE, and HMI science teams.

to Shallow sunspot model [5] , which is the prevailing sunspot model. 5 Mm is the largest stable sunspot with $B_o \approx 2.6 - 2.7$ kG according to the model. In this article we will show multiple results regarding the sunspot, and the radio flares and and other types of activity it produced, at various radio frequencies measured with MRO radio telescopes. We will compare the results to previous studies presented in [6] and [7] and show that NOAA 12192's activity was exceptional as it produced several flares during the ten days of its visibility.

INSTRUMENTATION

Solar maps at 37 GHz are produced with the 13.7-meter radio telescope, RT-14, which is a cassegain type antenna. RT-14 is located at Metsähovi Radio Observatory (MRO), Aalto University (Helsinki Region, Finland, GPS: N 60 13.04 E 24 23.35). The radio telescope provides full disk solar mapping, partial solar mapping, and, additionally, the ability to track any selected point on the solar disk. The beam size of the telescope is 2.4 arcmin at 37 GHz. The receiver is a Dicke type radiometer, thus, the radiometer's own noise will be filtered out. For the temperature stabilization of the receiver, a Peltier element is used. The noise temperature of the 37 GHz receiver is around 280 K, and the temporal resolution during the observations is 0.1 s or less. The radio emission comes from the solar chromosphere (transition region), and is recorded as intensity. More information about the Metsähovi RT-14 can be found in [7] and [8] .

Metsähovi RT-1.8 is a radio telescope with a 1.8 m dish diameter dedicated for continuous solar observations. The telescope has a beam size of 81.6 arc min and its system noise temperature is 270 K. It observes the total radiation of the Sun at a frequency of 11.2 GHz. The emission measured at 11.2 GHz originates from lower corona. The Quiet Sun Level (QSL) is around 12,000 K at 11.2 GHz. The radiotelescope is used for observing solar radio bursts, as it acts as a detector of general solar activity. Also studies on solar oscillations have been done. High sampling rate (5 kHz) enables studying flare fine structure, including short periodic oscillation phenomena. The radio telescope has no protective radome, therefore it is vulnerable to prevailing weather conditions. Around 200 solar radio bursts have been detected since its launch in 2001. Full documentation of the Metsähovi RT-1.8 can be found in [9] . The telescope has a logarithmic output which can be utilized in the case of strong bursts.

MRO joined the worldwide solar radio burst observing network, e-Callisto, in the fall of 2010 [10]. The network is coordinated by ETH Zürich and it consists of 69 instruments in 38 locations all over the world, covering solar radio spectral observations 24 hours a day throughout the year [11]. The network focuses on the decimeter and meter wavelength range. The first radio emission coming from the solar corona observed with the Callisto system was measured on September 24, 2010 [10]. The current frequency coverage in Metsähovi is 50 - 1450 MHz (the lower band is 50 - 846 MHz and the higher band 790 - 1447.250 MHz). Due to the low instrumental sensitivity studies of the quiet Sun are not possible. The changing radio emission indicates variation of solar magnetic activity. The birth of radio emission is affected by all plasma parameters, for example, temperature and density. This signifies the importance of radio observations and introduces interesting information on the subject. Gyromotion of thermal electrons in a presence of a magnetic field causes thermal gyroresonance emission. In active regions the magnetic field may have the strength to render the corona optically thick to enable gyroresonance absorption at the frequency range of 1 - 18 GHz. Plasma emission is the most prevailing emission mechanism in solar flares up to 3 GHz [12]. At mm- wavelengths gyro-resonance emission above the sunspot [13] and free-free radiation from plasma in active region loops are found [14].

OBSERVATIONS

Solar Radio Maps at 37 GHz
Both variations in brightness and in size of the radio brightening can be derived from the MRO 37 GHz observations. It should be noted, that the radio brightening located at the same location as the sunspot NOAA 12192. The 37 GHz results are described in the two following subsections.

Variation in Brightness
Variation in brightness temperature of the radio brightening is listed in the Table 1. The largest values are measured on October 26, 2014, when the brightness temperature is close to 200% compared to the Quiet Sun Level (QSL), which is 7800 K. On other days within the duration of the radio brightenings activity the level is between about 120 and 140% compared to QSL.

Variation in Size
Four selected solar radio maps during October 20 - 29, 2014 when the NOAA 12192 sunspot was both active and visible, measured at 37 GHz are shown in Figure 2. It can clearly be seen that the size of the radio brightenings varies with about the factor of 2.

Table 1: Brightness temperature, in per cents compared to QSL, of the radio brightening measured at 37 GHz from solar maps. Duration of each solar map scan is 7 min.

Date	Start time	Temperature	Date	Start time	Temperature
20.10.2014	09:50:24	103.23	25.10.2014	08:59:51	137.69
20.10.2014	10:00:49	124.06	25.10.2014	09:27:19	134.81
20.10.2014	10:10:48	122.55	25.10.2014	12:16:33	125.48
21.10.2014	08:40:29	119.81	26.10.2014	10:28:44	149.03
22.10.2014	08:03:19	125.37	26.10.2014	10:59:57	199.07
22.10.2014	08:21:49	125.83	26.10.2014	11:12:50	188.81
22.10.2014	11:10:11	122.64	26.10.2014	11:22:48	176.08
22.10.2014	12:04:49	121.93	26.10.2014	11:32:13	174.24
23.10.2014	10:07:36	120.79	26.10.2014	11:49:31	188.87
23.10.2014	10:18:53	119.78	28.10.2014	10:20:27	126.04
23.10.2014	12:22:47	118.03	28.10.2014	10:28:49	128.41
24.10.2014	07:31:01	121.77	28.10.2014	11:55:21	128.06
24.10.2014	07:40:02	144.28	29.10.2014	09:16:01	128.06
24.10.2014	07:50:10	145.07	29.10.2014	09:24:23	127.44
24.10.2014	07:58:58	124.88	29.10.2014	10:48:32	118.22
24.10.2014	10:43:27	124.43	29.10.2014	10:56:53	120.40
24.10.2014	11:42:48	125.62			

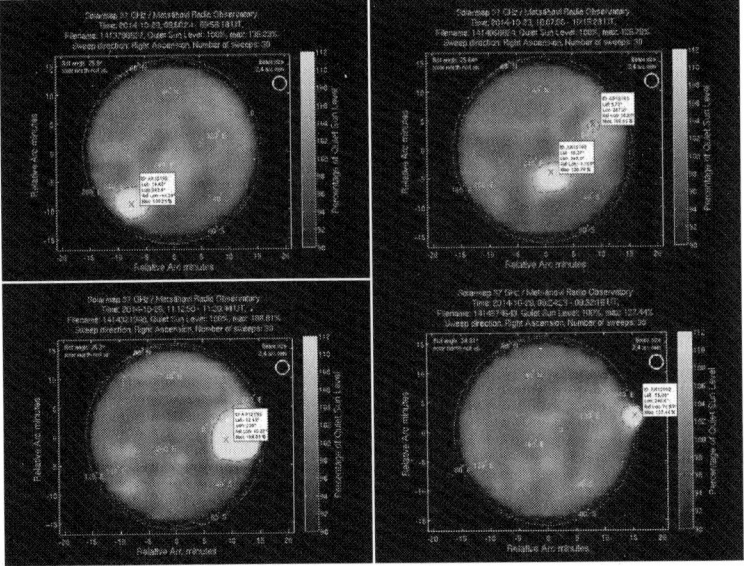

Figure 2. Solar radio maps measured on October 20, 23, 26 and 29, 2014.

Measurements at 11.2 GHz

We have listed all flares that exceed 100 Solar Flux Units (SFU) in Table 2. Peak flux values are estimations with errors approximately ±10%. An example of a solar radio burst measured with RT-1.8 on October 24, 2014 is in Figure 3. Observed solar radio bursts are associated with certain solar flares listed in Table 2. According to [15] about 20% of all X-class flares have strength of X2.0 - X2.9. When we take into account all possible flares, only less than 1% are X-class flares. The strongest of the now observed radio bursts at 2100 SFU is also rare in the long time series measured at MRO [6] . However, the previous time series did not include calibrated flux values.

Measurements with Callisto

In total, eight type III solar radio bursts listed in Table 3 were recorded at 200 - 400 MHz band of the e-Callisto. Also some events were recorded at lower frequencies, which are not included in this study. An example of a Callisto recording measured on October 24, 2014 is shown in Figure 4. The radio bursts were also observed with several other e-Callisto stations all around the world [11] .

Comparison with Long Time Series Measured at MRO

In comparison in [6], for example, in August, 2008, which was the most active month within the time series, 22 solar radio flares were measured. During other months in the time series, from January, 2000 to January, 2014, the number of flares exceeded ten only six times at the integration time of a month. During the ten days in October, 2014, 12 flares were measured at 11.2 GHz.

We also compared to the long time series study in [7], which includes data from 1978 to 2011 (solar cycles 21-14). In the study the strongest solar radio brightenings are 145% to QSL (11300 K). During October 26,

Table 2: Estimated maximum peak flux values at 11.2 GHz. Total of 12 flares were recorded at 11.2 GHz.

Date	Time	Estimated peak flux (SFU)	Solar flare classification [4]
20.10.2014	06:36 - 06:48	150	C6.3
20.10.2014	09:03 - 09:14	1600	M3.9
21.10.2014	10:28 - 10:37	180	
21.10.2014	10:48 - 10:50	170	
21.10.2014	12:27 - 12:29	340	C4.4
21.10.2014	13:17 - 13:20	110	
21.10.2014	13:37 - 13:38	940	M1.2
23.10.2014	09:44 - 09:52	510	M1.1
24.10.2014	07:40 - 07:46	1600	M4.0
24.10.2014	08:02 - 08:03	170	
26.10.2014	10:34 - 12:30	2100	X2.0
27.10.2014	09:20 - 11:00	no calibration available	M6.7

Table 3: Solar radio bursts, including the burst type, measured with the Callisto system (200 - 400 MHz).

Date	Time	Type
21.10.2014	10:01:11 - 10:02:13	III
21.10.2014	10:48:01 - 10:49:45	III
21.10.2014	13:36:02 - 13:37:57	III
23.10.2014	11:43:48 - 11:44:00	III
24.10.2014	07:39:34 - 07:41:43	III
24.10.2014	08:02:04 - 08:04:35	III
24.10.2014	10:01:36 - 10:02:05	III
26.10.2014	10:57:37 - 10:58:05	III

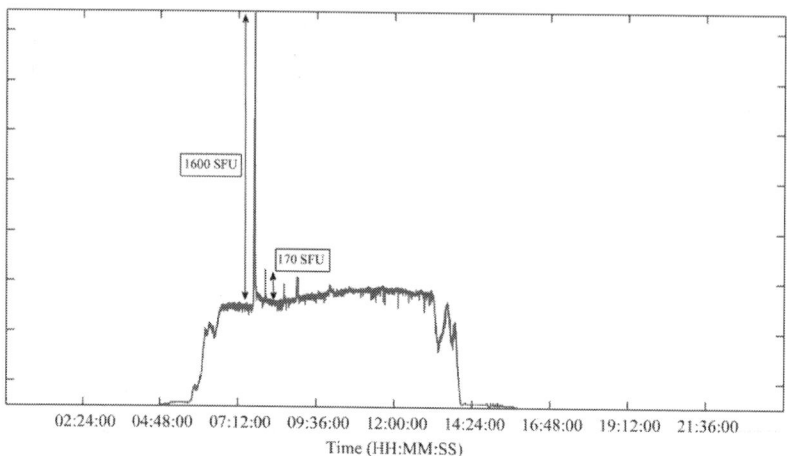

Figure 3: Two solar radio bursts on October 24, 2014 were measured at 11.2 GHz. The last wide spike is due to the calibration of the system.

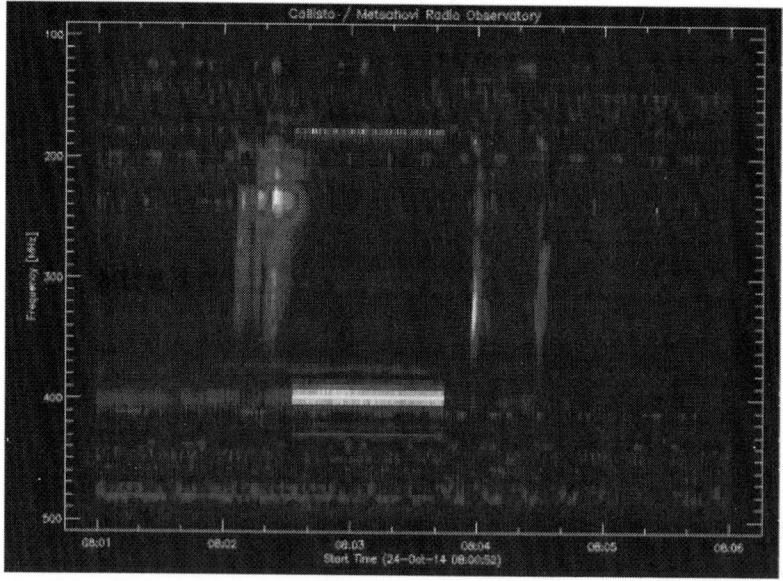

Figure 4: Type III solar radio burst measured with the Callisto system at 200 - 500 MHz on October 24, 2014. The large horizontal anomaly at about 400 MHz is due to RFI.

2014 145% QSL limit was exceeded on six occasions, and on its strongest point it reached about 200% QSL. The now recorded radio brightening belongs to the strongest category including less than 5% of radio brightenings ever measured at MRO.

CONCLUSIONS

In this article we have shown a review of radio signatures of NOAA 12192 measured at MRO at various instruments and radio frequencies. The radio brightening is remarkable even when the long time series measured at MRO are considered. The number of flares measured with RT-1.8 was 12 during the active period of the brightening, October 20 - 29, 2014, which makes this brightening one of the most active during the long time series ever measured at MRO. In comparison in [6], for example, in August, 2008, which was the most active month within the time series, 22 solar radio flares were measured. When the results were compared to the long time series [7] , it can be concluded that the now recorded radio brightening belongs to the strongest category including less than 5% of radio brightenings ever measured at MRO.

This exceptional sunspot will lead to follow-up studies. For example, several of the events measured at 11.2 GHz revealed interesting double-peak structures. Further studies of the double-peak structures will be conducted by the authors in the near future.

ACKNOWLEDGEMENTS

NOAA 12192 image created using the ESA and NASA funded Helioviewer Project.

REFERENCES

1. Kiess, C., Rezaei, R. and Schmidt, W. (2014) Properties of sunspot umbrae observed in cycle 24. Astronomy and Astrophysics, 565, Article ID: AA52.http://dx.doi.org/10.1051/0004-6361/201321119

2. Schad, T.A. and Penn, M.J. (2010) Structural Invariance of Sunspot Umbrae over the Solar Cycle: 1993-2004. Solar Physics, 262, 19-33. http://dx.doi.org/10.1007/s11207-009-9493-8

3. Zharkov, S., Zharkova, V.V. and Ipson, S.S. (2005) Statistical Properties of Sunspots in 1996-2004: I. Detection, North- South Asymmetry and Area Distribution. Solar Physics, 228, 377-397. http://dx.doi.org/10.1007/s11207-005-5005-7

4. NOAA's Space Weather Prediction Center's Web Site. http://www.swpc.noaa.gov/

5. Solov'ev, A. and Kirichek, E. (2014) Basic Properties of Sunspots: Equilibrium, Stability and Long-Term Eigen Oscillations. Astrophysics and Space Science, 352, 23-42.http://dx.doi.org/10.1007/s10509-014-1881-3

6. Kallunki, J. and Uunila, M. (2014) Total Solar Flux Intensity at 11.2 GHz as an Indicator of Solar Activity and Cyclicity. International Journal of Astronomy and Astrophysics, 4, 437-444. http://dx.doi.org/10.4236/ijaa.2014.43039

7. Kallunki, J., Lavonen, N., Järvelä, E. and Uunila, M. (2012) A Study of Long-Term Solar Activity at 37 Ghz. Baltic Astronomy, 21, 255-262.

8. Kallunki, J. (2013) Studies of Solar Activity with Emphasis on Quasi-Periodic Oscillations. Doctoral Thesis, University of Turku, Ser. A I Tom. 467, Turku.

9. Kallunki, J. (2009) Possibilities of the MetsähoviRadiotelescopes for Solar Observations. Licenciate Thesis, Faculty of Information and Natural Sciences, Helsinki University of Technology, Helsinki.

10. Kallunki, J., Uunila, M. and Monstein, C. (2013) Callisto Radio Spectrometer for Observing The Sun—Metsähovi Radio Observatory Joins the Worldwide Observing Network. IEEE Aerospace and Electronic Systems Magazine, 28, 5-9.http://dx.doi.org/10.1109/MAES.2013.6575404

11. e-Callisto Web Site (2014) International Network of Solar Radio Spectrometers.http://www.e-callisto.org/

12. Benz, A. (2006) Radio Emission of Solar Flare Particle Acceleration. 6th International Workshop on Planetary and Solar Radio Emissions, Graz, 20-22 April 2005, 325-338.

13. Shibasaki, K., Enome, S., Nakajima, H., et al. (1994) A Purely Polarized S-Component at 17-GHz. Publications of the Astronomical Society of Japan, 46, L17-L20.

14. Shibasaki, K., Alissandrakis, C.E. and Pohjolainen, S. (2011) Radio Emission of the Quiet Sun and Active Regions (Invited Review). Solar Physics, 273, 309-337.http://dx.doi.org/10.1007/s11207-011-9788-4

15. Le, G., Yang, X., Liu, Y., et al. (2014) Statistical Properties of X-Class Flares and Their Relationship with Super Active Regions during Solar Cycles 21-23. Publications of the Astronomical Society of Japan, 350, 443-447

CITATION

Uunila, M. and Kallunki, J. (2014) Radio Signatures of Sunspot NOAA 12192. *International Journal of Astronomy and Astrophysics*, **4**, 649-655. doi: 10.4236/ijaa.2014.44059.

Chapter 2

Doing Astronomy with Small Telescopes

KangujamYugindro Singh[1], IromAblu Meitei[2], Salam Ajitkumar Singh[1]

[1]Department of Physics, Manipur University, Canchipur, Imphal, India
[2]Department of Physics, Pettigrew College, Ukhrul, India

ABSTRACT

We are playing a lead role for growth of astronomy and its quality teaching and research in Manipur, a State located at northeast India (longitude = 93°58'E; latitude = 24°44'N; altitude = 782 m). We have innovatively designed and constructed three cost effective observatories, each costing a few hundred USD. These observatories are completely different in design and are perfectly usable for doing serious work on astronomical observation and measurements, using small ground-based telescopes. One Celestron CGE1400 telescope is housed with equatorial mounting in one of three constructed observatories and the same observatory has been inducted, since January 2012, as one of the members of the "Orion Project", which is an international project headquartered at Phoenix, Arizona, USA, dedicated for photometric and spectroscopic observations of five bright variable stars of the Orion constellation. We have been producing high precision BVRI photometric data that match well with those produced by other observatories enrolled in the Orion project. Our photometric data were presented and discussed in the 33rd Annual Conference of the Society for Astronomical Sciences: Symposium on Telescope Science, held at Ontario, California, USA during June 12 - 14, 2014. Further, we could successfully demonstrate them to the entire population of the State and play live shows of the observation of three spectacular astronomical events namely, solar eclipse of 15th January 2010, lunar eclipse of 10th December 2011 and Transit of Venus of June 6, 2012. We have conducted a number of seminars and workshops for training and research in astronomy. In the present paper, we would like describe our self-built observatories, our observational facilities, the BVRI photometric data that we acquired for the Orion project, and other activities undertaken for growth of astronomy activities in the State of Manipur, India.

INTRODUCTION

Having an observatory equipped with at least a telescope and starlight-measuring equipment such as photometer, CCD camera, spectrograph, etc. is first and foremost requirement for doing serious work in astronomy. We have three small telescopes (Celestron 9.25", Meade 12" and Celestron 14") and many backend telescope accessories including photometer, CCD camera, spectrograph, etc., purchased at different times during 2004 to 2012. However, because of financial constraints we could not afford to purchase commercially available ready-to-use observatories such as the Sirius observatory made in Australia, which would cost about USD 25,000 (inclusive of shipping, custom duty and other charges) on delivery at our place. To meet our requirement for housing our telescopes, we have innovatively designed three observatories and constructed them, each costing a few hundred USD. These observatories are completely different in design and are perfectly usable for doing serious work on astronomical observations and measurements. Out of the three observatories, the one which houses the Celestron 14" telescope has been put to use, since January 2012, for contributing BVRI photometric data to the international "Orion Project" headquartered at Phoenix, Arizona, USA, which is dedicated for photometric and spectroscopic observations of five bright variable stars of the Orion constellation [1] . Our photometric data were presented and discussed in the 33rd Annual Conference of the Society for Astronomical Sciences: Symposium on Telescope Science held at Ontario, California,

USA during June 12-14, 2014 [2] . Our CCD cameras have been tested by taking CCD images of the Moon, Jupiter and Saturn and our spectrographs also have been tested with standard light sources. They are ready to be used as integrated components of our observatories.

Because of our strong enthusiasm and commitment for astronomy popularization and education in our State, Manipur (longitude = 93°58'E; latitude = 24°44'N; altitude = 782 m), we could successfully demonstrate them to the entire population of the State and play live shows of the observation of three spectacular astronomical events namely, Solar eclipse of 15th January 2010, Lunar eclipse of 10th December 2011 and Transit of Venus of June 6, 2012. Besides these, we have conducted a number of seminars and workshops for training and research in astronomy. We are also playing a lead role in the State for growth of astronomy and its quality teaching and research in the University and College sectors.

In what follows we innovatively describe our self-built observatories (in Section 2), activities for the international Orion project (in Section 3), measures taken up for astronomy education (in Section 4) and a brief conclusion (in Section5).

SELF-BUILT OBSERVATORIES

We have built three cost effective observatories in different models for housing our 3 telescopes (Celestron 9.25", Meade 12" and Celestron 14"). All of these observatories are now fully functional.

Figure 1 is the picture of a self-built dome type observatory used for housing one Celestron 9.25" telescope. This observatory was constructed in 2009 at a cost of about 700 USD. The unique feature of this observatory is that it has a hemispherical roof with a collapsible strip, which can rotate along a circular groove running on the top of the main cylindrical body. Opening the strip and rotating the dome can make the entire sky accessible to the telescope housed inside the structure.Table 1 gives the physical parameters of this observatory.

The Celestron 9.25" telescope housed in this observatory has been used, after an accurate equatorial mounting, for planetary imaging using Optec's ST7 CCD camera and for photometric observations of some variable stars using Optec's SSP3A photometer. A weakness of this observatory is that a single person cannot manage controlling both the telescope and the upper movable doom; it requires at least two persons, one for controlling the position of the movable upper doom and the other for controlling the telescope and taking observations.

Figure 1. Picture on the left is a self-built dome-type observatory and that on the right is the in-housed Celetron 9.25" telescope.

Table 1. Physical parameters of the dome type observatory (Figure 1)

Diameter of dome at the base	7.5'
Height of the cylindrical body	4.5'
Breadth of sky opening stripe	1.5'
Max dome height from the base	7'3"
Area of entrance door	4.5' × 2'10"

Figure 2 is the picture of a self-built hut-type observatory used for housing one Celestron 14" telescope. This observatory, which was built in 2011 at a cost of just USD 200, is typically hut-type with a tin roof. It has a main entrance door. The structure as a whole is movable with 4 pairs of wheels: 2 pairs in the front and 2 pairs in the back. Keeping the main door open, the structure as a whole can be taken away from in-housed fixed telescope so that the entire sky is accessible to the telescope. Table 2 gives the physical parameters of this observatory.

The Celestron 14" telescope housed in this observatory has been in service, also after an accurate equatorial mounting, for taking photometric observation of five Orion variable stars, as a part of the international "Orion Project" described in the next section i.e., Section 3, using an Optec's SSP3 photometer. One SBIG made DSS7 coupled to an Optec's ST7 camera are going to be inducted soon for this observatory for taking part in spectroscopic observation of the five Orion stars, also as part of the Orion project. This hut-type observatory is found to be easy and convenient for taking astronomical observations; a single person can manage the whole observation session.

Figure 3 is the picture of a self-built bullet-type observatory used for housing one Meade 12" telescope. This observatory which was built in April, 2014 at a cost of about USD 200 is typically a bullet-type with a tin roof head. Its lower portion is fixed while the upper portion i.e., the tin roof head, is movable on pair of rail tracks by means 4 wheels fixed on the base foot of the roof head. It has an entrance door. By moving away the upper portion, the telescope can see the whole sky. Table 3 gives the physical parameters of this observatory.

The Meade 12" telescope housed in this observatory is permanently mounted on an equatorial wedge and it will be used for CCD photometry of some SU UMa stars i.e. a subclass of dwarf novae, for looking out their bursts, super outbursts and super humps which are characteristics of the star systems. A single person can easily control the observatory and conduct an observation session.

Though all of the three self-built observatories described above, are usable for scientific research in astronomy, the hut-type observatory shown in Figure 2 and the bullet-type observatory shown inFigure 3 are more practical in the sense that they are very cost effective and easy to operate. Therefore, the hut-type and bullet-type observatories may be good examples for all researchers including amateur astronomers who are constrained with limited budget but wish to make meaningful contribution in scientific research in astronomy, with small telescopes. Potential fields of research which can be done with small telescopes housed in such proto-type observatories include astrometry, imaging, photometry, polarimetry and spectroscopy. For producing high-precision standard data amateurs should keep in touch with global organizations such as American Association of Variable Star Observers (AAVSO), Society for Astronomical Sciences (SAS), French Association of Variable-Star Observers (AFOEV), Centre for Backyard Astrophysics (CBA), etc. Such a standard data can definitely supplement the results from large observatories and will be worthwhile for doing science in Astronomy.

THE INTERNATIONAL "ORION PROJECT"

Since January 2012, MU observatory has been participating, as the lone member from India, in the International "Orion Project" [3] , headquartered at Phoenix, Arizona, USA for photometric and spectroscopic observation of five variable starsof the Orion constellation namely, Betelgeuse (alpha Orionis), Rigel (beta Orionis), Mintaka (delta Orionis), Alnilam (epsilon Orionis) and Alnitak (zeta Orionis). As on September 2, 2014, eighteen observatories across the globe (twelve from USA, one each from India, UK, France, New Zealand, Austria, Germany) are taking part in the Orion project. The details of the eighteen observatories are given Table 4.

The Orion project has several goals. First goal is to help beginners with photometry and spectroscopy. The second goal is the actual observations and data acquisition. The project stems out on realising that the programstars, because of their intense brightness, are seldom observed in detail and that they pose a serious problem for observation in most professional observatories. The project speculates that by having these stars observed over a long time, interesting changes might be seen that will warrant closer investigation. The third goal of the project is to generate an excellent archive of photometric and spectroscopic data.

Figure 2. Self-built hut-type observatory (left) and the in- housed Celestron 14'' telescope (right).

Figure 3. Self-built bullet-type observatory (left) and the in-housed Meade 12'' telescope (right).

Table 2. Physical parameters of the hut-type observatory (Figure 2)

Base area of main body	4'2" × 7'
Height of main body	7'
Height of roof top from the base	9'6"
Area of entrance door	4'2" × 7'

Table 3. Physical parameters of the bullet-type observatory (Figure 3)

Base area of the fixed portion	6'7" × 5'3"
Height of the fixed portion	5'1"
Height of the roof from the base	7'1"
Area of entrance door	4'2" × 2'

The authors of this paper are the observer members of MU observatory, India. A photo snap of these authors taken along with the equipment used for the Orion project is shown in Figure 4.

The photometric data obtained from MU observatory for the Orion project were presented in the 33rd Annual Conference of the Society for Astronomical Sciences: Symposium on Telescope Science, held at Ontario, California, USA during June 12-14, 2014 [4] .Meissa was used as the comparison star. The BVRI band photometry obtained by using SSP3 photometer, of the five stars of the Orion project under study namely, Betelgeuse, Rigel, Alnilam, Alnitak, Mintaka are shown in Figures 5-9. The data points marked with "MU" are due to MU observatory and those marked with "SKED" are due to Sikes Observatory, Arizona, USA [5] .

Table 4. Observers of the Orion project

Observer	OC/AAVSO ID	Observatory	Location	Telescope	Spectrograph/Camera	Photometer
Jeff Hopkins	HPO	Hopkins Phoenix Observatory	Phoenix, Arizona USA	12" LX200 GPS SCT	Lhires III 2400/600 l/mm; ALPY 600; Star Analyser	SSP-4
Gene Lucas	NOAO	Night Owl Astrophysical Observatory	Fountain Hills, Arizona USA	10" LX200 SCT	Orion Starshoot G3; Star Analyser 100 l/mm; DSI Pro	BVRI DSI Pro
Joe Gianninoto	DDO	Devil Dog Observatory	Tucson, Arizona USA	NP 127is, Mach AP GTO	Star Analyser 100 l/mm	N/A
Laurent Corp	CLZ	Garden Observatory	Rodez, France	8" Newtonian F/5.92 Fork Mount	N/A	CCD ST7 BVRI
Yugindro, Singh & I. Ablu Meitei	MU	Manipur University	Manipur, India	Celestron 14 CGE 1400 XLT	N/A	SSP-3 BV
Ken Sikes	SKED	Sikes Obervatory	Chandler, Arizona USA	10" LX SCT	N/A	SSP-3a
Jim Tubbs	TJDA	Tubbs Observatory	Twin Falls, Idaho USA	8" LX200 GPS F/6.3	Star Analyser 100 l/mm	N/A
Carl Knight	KCD	Ngileah Observatory	Bulls, New Zealand	12" LX90 LNT F/10	N/A	SSP-4JH
Steve Spears	SSCC	Spears Observatory	Westlake, Ohio USA		Star Analyser 100 l/mm	N/A
Phil Hoppes	HPEA	Ovegaard Outreach Observatory	Overgaard, Arizona USA	10" LX200 Classic	L200 1200/600/300 l/mm; Atik 314+	CCD UBVRc, lc
Steve Cuthbert	SCO	Cuthbert Observatory	York, United Kingdom	Espirit 120 mm f/7	ALPY 600; ATIK 314L+	
John Menke	JMO	John Menke Observatory	Barnesville, Maryland USA	C11. Newtonian 18" f/3.5	DSS7; ST402, ST1603	N/A
Jerry Persha	PGD	Jerry Persha Observatory	Lowell, Michigan USA	Meade 10" SCT Classic, F/10	N/A	SSP-4 JH
Al Stiewing	ASO	Al Stiewing Observartoy	Phoenix, Arizona USA	8" SCT. C-8	N/A	SSP-4 JH
Dr. Thomas Schroefl	ANO-VST	Austrian North Observatory-Very Small Telescope	Vienna, Austria	11" C-11	Star Analyser; DADOS 200/900/1200 l/mm	N/A

ASTRONOMY EDUCATION

Besides our participation in the Orion project, we have also conducted a number of programmes, using small telescopes, for popularization and education of astronomy amongst the public. We could demonstrate successfully to the entire population of the State, live shows of the observation of three spectacular astronomical events namely, Solar eclipse of 15th January 2010, Lunar eclipse of 10th December 2011 and Transit of Venus of June 6, 2012. The observation of these events were widely publicised and covered by print and electronic media. The photo in Figure 10 shows K. Y. Singh having a live TV phone-in-iteraction program with the public on observation of the total lunar eclipse on 10th December, 2011, through a State Television Network (ISTV). A Celestron CGE925 telescope was used for observation of the total lunar eclipse.

Our observation of the Transit of Venus on June 6, 2012 was spectaculary successful as thousands of people participated in observation of the event. One Meade 12 inch telescope fitted with a Coronado SolarMax II 90 mm Hydrogen alpha filter and a DSLR Camera was used for observation of the transit. Photographs and video recordings were also taken with the help of a DSLR camera fitted to the telescope. Some photo snaps and a video recording of the observation of the Transit of Venus were published in the YouTube networks [6] , [7] . A news report of the observation of the event is available at the network [8] .

Figure 4. A photo snap of K. Y. Singh (left) & I.A. Meitei (right) taken with Celestron CGE1400.

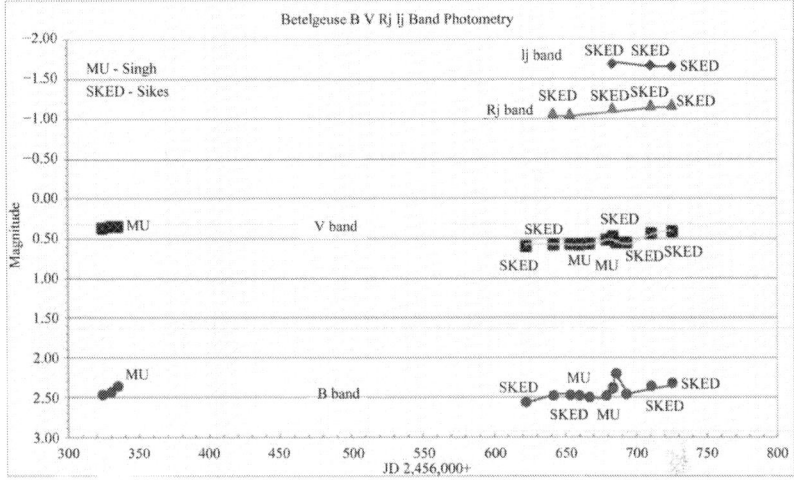

Figure 5. SSP-3 BVRI photometry of Betelgeuse.

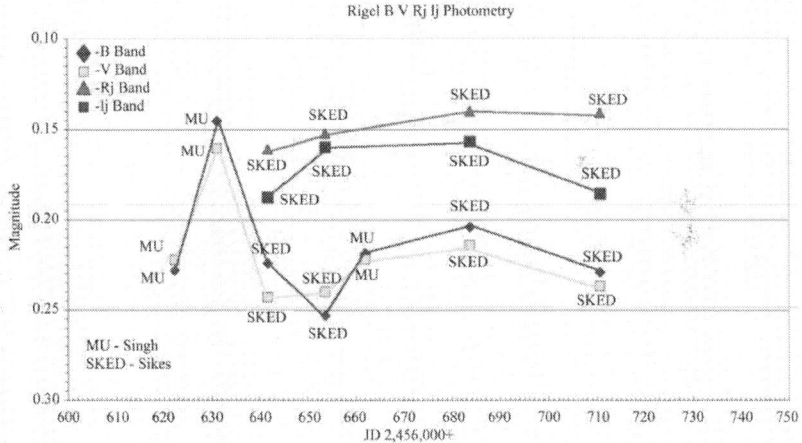

Figure 6. SSP-3 BVRI photometry of Rigel.

Figure 11 shows a photo snap of the Transit of Venus taken at 08:47:12 ISTon June 5, 2012.

Figure 12 and Figure 13 show two scenes where people including school children thronged at the observation site to observe the transit directly with their own eyes.

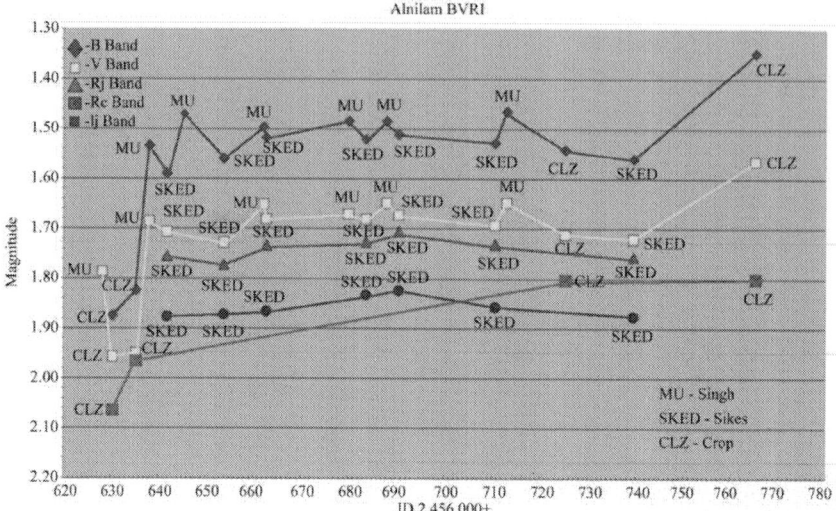

Figure 7. SSP-3 BVRI photometry of Alnilam.

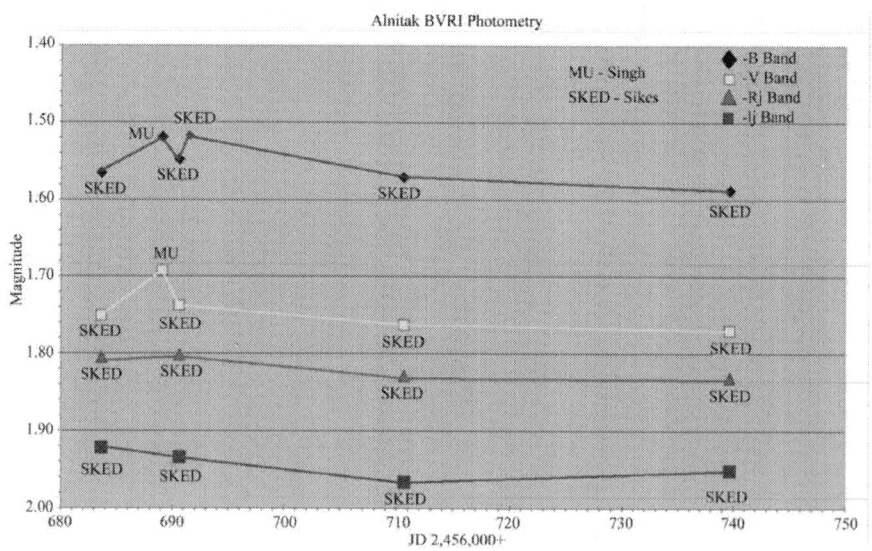

Figure 8. SSP-3 BVRI photometry of Alnitak.

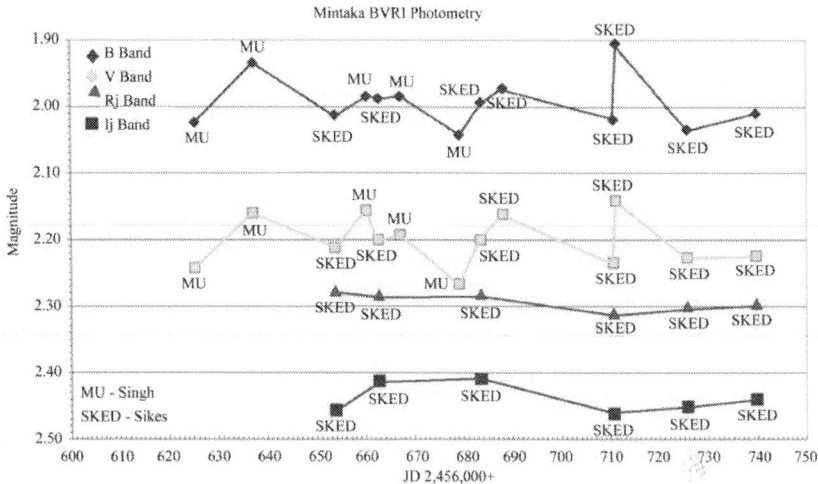

Figure 9. SSP-3 BVRI photometry of Mintaka.

One of the authors of this paper namely, K. Y. Singh has been playing a key role for growth of astronomy activities in Manipur University, where he is the lone faculty member teaching Astronomy, since 2009, in the Department of Physics of the University; he has strong enthusiasm and commitment for astronomy education. Under his cordinatorship, a number of seminars and workshops have been organized at Manipur University, the latest one being the 3-Day IUCAA sponsored "Introductory Workshop in Astronomy & Astrophysics" which was held during February 10-12, 2014. The purpose of the latest workshop was to expose and motivate post- graduate students, young researchers and college/university teachers towards Astronomy and Astrophysics. The workshop focused on observational astronomy based on small telescopes and its recent advances. A total of 37 participants from different parts of the country (India) participated in the workshop. Figure 14 shows a photo taken on the Inaugural Function of an IUCAA sponsored "Introductory Workshop in Astronomy & Astrophysics" held at Manipur University, India during Feb 10-12, 2014 under coordinatorship of K. Y. Singh.

Figure 10. Live TV phone-in-interaction program of K.Y. Singh (left) with public regarding the total lunar eclipse on 10th Dec, 2011.

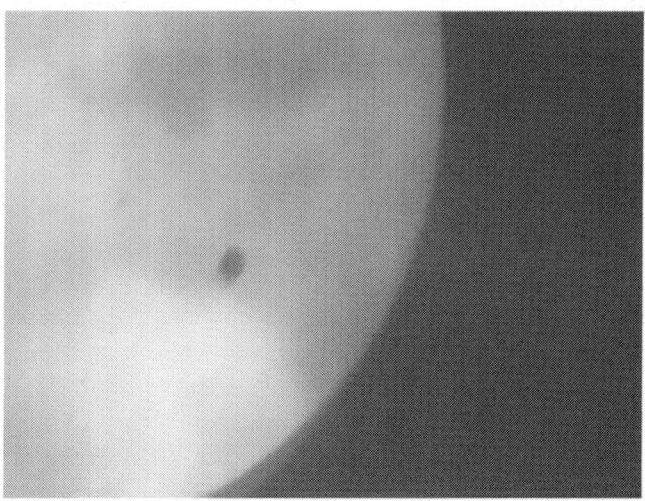

Figure 11. Position of Venus observed at 08:47:12 ISTon June 5, 2012; red spot seen on the sun disc is the shadow of Venus.

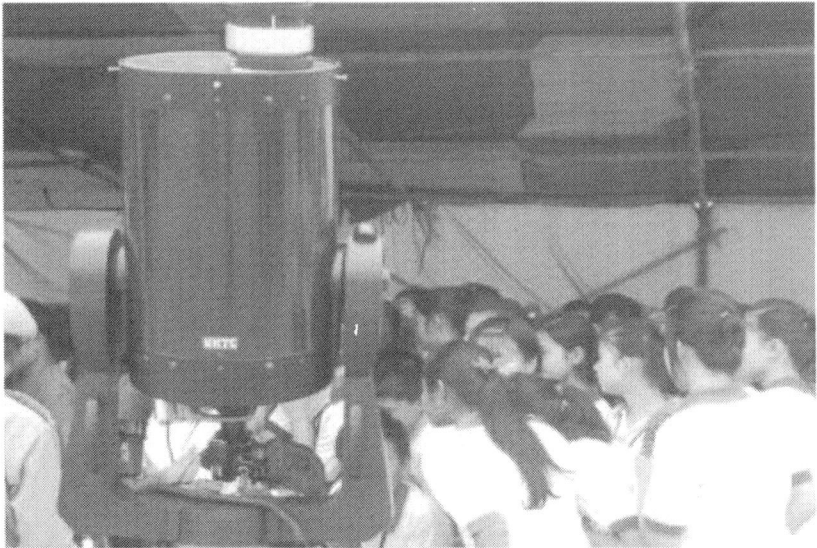

Figure 12. A scene where people were in queue for observing the transit of Venus on June 6, 2012.

Figure 13. A scene where school children rushed to have a glimpse of the transit of Venus on June 6, 2012.

Figure 14. Inaugural function of a national workshop held at Manipur University, India during Feb 10-12, 2014.

A report of the workshop was published in Khagol (A quarterly bulletin of the Inter-University Centre for Astronomy and Astrophysics (IUCAA), Pune, India) in its issue no. 98, April 2014 and is available at the web link [9] . Besides teaching Astronomy at the post-graduate level, implementing many government funded research projects, producing many Ph.D. degrees and presenting many invited talks in conferences, seminars and workshops, the astrophysics team led by K. Y. Singh has published a number of research papers in many international journals of repute, thereby reflecting fast growth of astronomy activities in Manipur, India.

CONCLUSION

We have innovatively designed and constructed three cost effective observatories of different models for housing small telescopes. These observatories can be used for scientific research and education and will be a good example to other researchers and institutions with limited budget. Using one of the observatories, we have been an active partner of the international "Orion Project" and have been producing SSP3 VBRI photometric data for the project. Using our observational facilities, we have done substantial works for popularization, education and research in astronomy in Manipur, a State in the northeast part of India. The presence of the observatories has brought fast growth of our astronomy activities with strong commitments.

ACKNOWLEDGEMENTS

S. Ajitkumar Singh is grateful to the Indian Space Research Organization, Department of Space, Government of India for providing him a JRF under a RESPOND Project.

REFERENCES

1. The Headquarter of the International Orion Project.http://www.hposoft. com/Orion/Orion.html
2. Society for Astronomical Sciences, USA. http://www.socastrosci.org/ symposium.html
3. Observers of the Orion Project. http://www.hposoft.com/Orion/ Observers.html
4. Hopkins, J.I. (2014) Orion Project: A Photometry and Spectroscopy Project for Small Observatories. Proceedings for the 33rd Annual Conference of the Society for Astronomical Sciences: Symposium on Telescope Science, 12-14 June 2014, Ontario, 93-104.
5. Sikes Observatory, Arizona, USA. http://www.hposoft.com/Orion/ Observers/SO.html
6. The Uploaded Photos of the Observation of the Transit of Venus on June 6, 2012, at Manipur University, India. http://www.youtube.com/watch?v= DJiTr4D3kLg
7. The Uploaded Video of the Observation of the Transit of Venus on June 6, 2012, at Manipur University, India. http://www.youtube.com/watch?v= MzthHnUIKZmM
8. A News Report on the Observation of the Transit of Venus on June 6, 2012, at Manipur University, India. http://e-pao.net/GP.asp?src=17..070612.jun12
9. (2014) Khagol Published by IUCAA, Pune, India. Issue No. 98, 7.http://ojs.iucaa.ernet.in/index.php/khagol/article/view/159/142

CITATION

Singh, K. , Meitei, I. and Singh, S. (2014) Doing Astronomy with Small Telescopes. International Journal of Astronomy and Astrophysics, 4, 560-570. doi: 10.4236/ijaa.2014.44052.

Chapter 3

Doppler Boosting May have Played no Significant Role in the Finding Surveys of Radio-Loud Quasars

Morley B. Bell

Herzberg Institute of Astrophysics, National Research Council, Ottawa, Canada

ABSTRACT

There appears to be a fundamental problem facing Active Galactic Nuclei (AGN) jet models that require highly relativistic ejection speeds and small jet viewing angles to explain the large apparent superluminal motions seen in so many of the radio-loud quasars with high redshift. When the data are looked at closely it is found that, assuming the core component is unboosted, only a small percentage of the observed radio frequency flux density from these sources can be Doppler boosted. If the core component is boosted the percentage of boosted to unboosted flux will be higher but will still be far from the 90 percent required for Doppler boosting to have played a significant role. Without a highly directed, Doppler boosted component that dominates the observed flux, radio sources found in low-frequency finding surveys cannot be preferentially selected with small jet viewing angles. The distribution of jet orientations will then follow the sini curve associated with a random distribution, where only a very few sources (~1%) will have the small viewing angles (<8°) required to explain apparent superluminal motions $v_{app} > 10c$, and this makes it difficult to explain how around 33% of the radio-loud AGNs with high redshift can exhibit such highly superluminal motions. When the boosted component is the dominant one it can be argued that in a flux limited sample only those members with small viewing angles would be picked up while those with larger viewing angles (the un-boosted ones) would be missed. However, this is not the case when the boosted component is small and a new model to explain the high apparent superluminal motions may be needed if the redshifts of high-redshift quasars are to remain entirely cosmological.

INTRODUCTION

Recently López-Corredoira [1] has reminded us that there are still many quasar/QSO observations that remain difficult to explain. Here we discuss what appears to be another of these. The large apparent superluminal motions observed in the jets of many radio-loud quasars can be explained by assuming either, 1) that the objects are at their cosmological redshift (CR) distance and almost all of their radio flux density comes from ejected material that is relativistically beamed towards us in a highly collimated jet at near light speed and with a small inclination angle, i, close to the line-of-sight [2,3]; or 2) that the objects are much closer than their redshifts imply so the observed angular motions in their jets lead to only subluminal linear speeds [4,5]. It has been claimed that the former model not only explains the apparent superluminal motions, but that it can also, through Doppler boosting, explain why most of the detected sources would naturally have very small inclination angles. However, for this model to work, one of its main requirements is that, in the finding survey, the Doppler boosted component of the source flux density must be the dominant one. Whether or not this requirement is met therefore needs to be examined closely. To do this we first examine what source material can be moving towards us at relativistic speeds in a tightly confined beam. We then consider what percentage of the total source flux density the radiation from this material contributes. It will be demonstrated below that with the existing observational evidence it may no longer be possible to use the relativistic beaming model to explain the high percentage of radio loud quasars exhibiting superluminal motion.

THE ROLE OF DOPPLER BOOSTING

The problem of explaining apparent superluminal motion in quasar jets was looked at closely over twenty years ago [6]. Since that time much new information has been obtained on the jets of many more radio loud quasars. Much of it [7,8] is of excellent quality, and some of it has resulted in movies being made that depict reasonably clearly what is taking place near the central engines of these objects when ejections occur. Unfortunately the lack of adequate resolution near the central compact object still prevents us from obtaining a clear picture of the jet launching process. If material is ejected from a source at relativistic speeds, because of Doppler boosting its radiation in the direction of motion will be enhanced, and radio finding surveys will preferentially pick up those sources that are ejecting material towards us [8]. This is only true, however, as long as the boosted component is the dominant one. The largest boosts in intensity occur for sources with jet viewing angles $i<$ $8°$ (see **Figure 20** and equation B5 of [9]). Thus, as pointed out above, if a high percentage of the sources show apparent superluminal motion in their jets it can be explained if most of the radiation has been Doppler boosted and comes from material whose ejection speed is relativistic and whose direction of motion is towards us (close to the line-of-sight). If none, or only a small percentage, of the radiation is Doppler boosted, and there are no other selection effects active, most of the sources would have been detected without the boosted component. The sources will then have a close to random distribution of orientations (sini) in which 50% will have jet viewing angles that are greater than $60°$, and only ~1% will have $i< 8°$. In this case, if a large percentage of the sources show large apparent superluminal motions, another way of explaining these motions must be found. Although they can be explained by bringing the sources closer to us until the linear speeds calculated from their angular motions are no longer superluminal, this argument has been found unacceptable because it requires that the redshifts of quasars contain an additional intrinsic component unrelated to the normal cosmological, or distance-related, one.

JET/COUNTERJET ASYMMETRY

It has been argued that the jet asymmetry, or one-sidedness, seen in many of these objects at 15 GHz, is a strong indication for Doppler boosting in the approaching jet. But recently, an attempt to show that the asymmetry in the jets of M87 at 15 GHz could be explained by relativistic motion gave negative results [10]. No evidence was found for relativistic ejections at 15 GHz in spite of the obvious jet/counterjet asymmetry. These authors were forced to conclude that the large jet/counterjet asymmetry in the inner jets of M87 may be intrinsic and not due to Doppler boosting. This was a significant result that appears to negate one of the main claims of the proponents of Doppler boosting; namely that a jet/counterjet asymmetry is evidence for Doppler boosting. Although relativistic motions have been claimed in the M87 inner jet at X-ray wavelengths [11], there is no way to be certain that it is the same material that is being observed at 15 GHz. In fact, it has been suggested that the X-ray event may represent an entirely different phenomenon (see **Figure 6** of [12]). The important point here is that when the same observing frequency (15 GHz) was used to observe both the jet motion and the asymmetry, only non-relativistic ejection speeds were seen in the material that showed asymmetry.

It would seem then that the jet one-sidedness seen in so many of these sources may originate simply because the strength of the jet is associated with the amount of material that is moving from the accretion disc to the jet at any given time, and that this material, associated with the flaring type of jet ejections discussed here, is normally only accreted from one side of the disc at a time. Unfortunately, the exact process by which material is ingested into the central object and regurgitated in the jet is still not well understood. However, the evidence clearly indicates that intrinsic asymmetry is common and its presence then cannot automatically be assumed to imply relativistic beaming in many sources.

WHAT PERCENTAGE OF THE RADIATION COMES FROM MATERIAL MOVING OUTWARD AT RELATIVISTIC SPEEDS?

We now need to determine what percentage of the total radio radiation from these sources is likely to originate in material that is moving away from the central object at relativisitic speeds, and in a tightly confined beam. To help in this examination the different emission regions found in jetted sources are shown in **Figure 1**. There are three main regions: 1) an extended, kpc-scale jet that is resolved in the VLA observations but lies outside the field of view of the VLBA when the core is included. Although not shown here, the extended jet may also include at its end a giant radio lobe with hot spots; 2) an inner, parsecscale jet that is well resolved in VLBA observations but unresolved by the VLA and; 3) a compact core component that is unresolved in all cases. In the original finding surveys that were carried out at long radio wavelengths it is entirely possible that much of the radiation from the inner, pc-scale jet may have originated from a region that was too deep inside the long-wavelength radio photosphere to have been detectable.

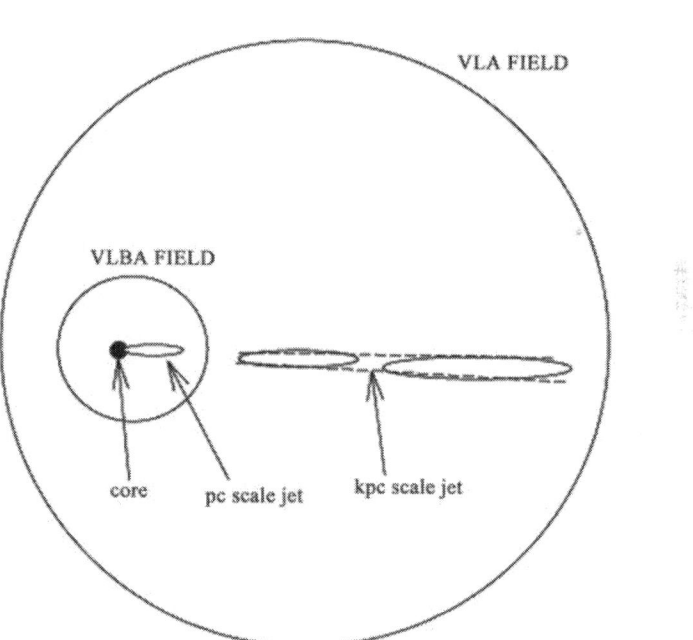

Figure 1. Showing relative jet areas covered by VLBA and VLA (not to scale). Because the core is unresolved a small portion of the inner, pc-scale jet flux will be included in the peak flux of the core component. However, because in most cases the inner jet covers several beam areas, this is expected to be small.

RADIATION FROM THE OUTER JET AND GIANT RADIO LOBES

As noted earlier, when the Doppler boosted component is small compared to the total flux density it cannot introduce a strong selection effect that will preferentially pick up sources with small inclination angles in the finding surveys in which most of these radio-loud AGN galaxies were discovered. Almost all radio-loud AGN galaxies (quasars, BLLacs) were found in the early radio surveys (Parkes, Cambridge 3C and 4C), that were carried out at low frequencies (178 or 408 MHz) with large antenna beamwidths. The beamwidth of the Parkes telescope at 408 MHz, for example, is 48 min of arc [13]. Consequently, the finding surveys would have detected the total radio radiation coming from these sources. This is especially true for high redshift sources where the largest angular size is less than ~3 min of arc for sources with z > 0.1 [14]. Even the 4C survey, which had a 1.3 min of arc beam, would have detected the total radiation from sources with z > 0.2, which includes most of the radio-loud quasars. The 4C detection limit was 2 Jy.

When a jetted source contains giant lobes with internal hotspots these features will almost certainly contain most of the source flux. Since outward motion in the lobes has been shown to be close to 0.02c ± 0.01c [15-17] the radiation from the lobes will be un-boosted.

Deceleration of the flow in the kpc-scale jet has also been examined by several previous investigators [17-22] and the flow is found to slow down quickly to near 0.1c beyond a few kpc from the core. As a result radiation from most of the kpc-scale jet must then also be un-boosted. Thus none of the radiation from the lobes and almost none of that from the kpc-scale jet can be included in the relativistic flow radiation component. This means that the boosted radiation component must come mainly from the core or inner pc-scale jet.

RADIATION FROM THE CORE AND INNER JET

The question of whether or not the radiation from the compact central core in core-dominant sources is boosted is obviously an important one. Because of this a lot of effort has been devoted to trying to prove that the core radiation, which is unresolved even with the mas resolution of the VLBA, is boosted in the jet direction. Some investigators have argued that the core component, although stationary, is actually part of the jet base [23-25], and that the lack of core motion in this case is because the material in this region is still being accelerated and has not yet reached relativistic speeds. But if this were the case it would not matter whether the core is associated with the accretion disc or the jet base, its radiation cannot be boosted if the radiating material is not moving relativistically.

Recently it was demonstrated [26] that most of the radio frequency radiation from the strong, unresolved cores of these objects could not originate in the jet, and must be coming from a separate region centered on the central compact object and accretion disc. It is apparent [26] from the investigation of 3C279 by [27] that there are three separate radiation components involved in producing the total radio radiation from the compact core and inner jet. These are as follows:

1) The first is a flaring component that only becomes visible when a new ejection event commences, and then only after the radiating material being ejected, or its shock front, passes beyond its relevant photosphere, which is the point beyond which the external medium is transparent to the wavelength being observed. This component is jet related and at radio frequencies represents only a small percentage of the total flux density observed (comparable to that from an individual blob seen after the ejected material begins to be resolved in the inner jet). For outward motion in the jet, since we see deeper at short wavelengths, the shorter wavelength flares (γ-ray, X-ray, optical) will appear before the radio flare. It is thus possible to estimate a lower limit to the separation between any two photospheres from the time delay in light days between the appearance of their respective flares. Recently, [28] have carried out an analysis of the flaring behavior of 3C454.3 using short wavelength data (optical, X-Ray, γ-ray), as well as mm-wave. They find a time delay of 30 ± 15 light days between the short wavelengths and the mm-wave flares. This corresponds to a distance of ~0.025 pc if the jet material is moving out relativistically. It is generally accepted that the high energy flaring radiation comes from the unresolved region very close to the accretion disc, and from the 3C454.3 results this appears to be confirmed, with the radius of the mm-wave photosphere likely

to be less than 0.1 pc. From these results it then seems likely that the radius of the radio photosphere lies well inside the half-power beam of the VLBA, even at 43 GHz. Note that in this model there is no Blazar zone in the jet of the type postulated by [29]. For a particular wavelength this zone is replaced in our model by the point in the envelope surrounding the accretion disc at which the jet first becomes visible (its relevant photosphere). This point is, unfortunately, still unresolved by present-day instruments. At radio frequencies this flaring component; is much weaker than the total core component; is jet related but still unresolved; and is expected to be boosted if the jet is pointing towards us. The visibility, or luminosity, of the flaring components increases towards shorter wavelengths and this can be explained by the decrease in radius of the respective photospheres which translates into an increase in the magnetic field strength closer to the central compact object.

2) The second component observed in the flux monitoring of 3C279 is one that can be referred to as the slowly varying component. It is easily shown that this component comes mainly from the inner jet, increasing and decreasing directly with the number of blobs present in the jet. This component is entirely related to the jet material, varies continuously, and if moving in our direction is expected to be Doppler boosted. In the most active sources this component can be as least as strong as the core component. Although the highly variable sources (like 3C279 and 3C454.3) are the most highly studied, the majority of core-dominant sources do not fall into this highly variable category and in most cases the steady core component is the dominant one.

3) The third component is a steady or non-varying one that contains most of the flux from the unresolved radio core. Since this radiation is detectable, at radio frequencies it must come from a radius larger than that of the radio photosphere. Even with the best resolution available this component is centered on the accretion disc, and shows no sign of motion. There is no evidence that any of the nonvarying core component is associated with the jet flow. However, because both it and the flaring component are unresolved, the two will be superimposed, even with the resolution provided by the VLBA.

It was demonstrated [26] that the steady, non-flaring core component cannot be explained by a continuous jet flow component, which, if it were part of the jet, would be needed to explain its steady nature.

Jorstad et al. [28] argue that the radio core lies at the end of the acceleration zone at the base of the jet. This core-in-jet model can easily be ruled out when there is no continuous flow, because this dominant, non-varying core component is still visible even when there is no ejection event taking place to be accelerated. Furthermore, if it were really part of the jet, it is not clear how this strong radio core-in-jet can show no sign of motion, while motion is readily seen as soon as the material moves outside the photosphere. If the radiation is coming from a region in the jet that is not moving, and produced by particles that are passing through it at relativistic speeds, the radiation from this stationary core-in-jet material still cannot be boosted. Is no motion seen in the acceleration region because the viewing angle is close to 0°, while motion is readily detected beyond the core-in-jet because there is a change in the jet direction? Although it is suggested in their core-in-jet model that there may be a sudden change in the jet direction at this point, the likelihood that every source would have this same bend seems small. It is also interesting that, while no proper motion is seen in the core, relativistic motion in the inner jet is readily detected even though the viewing angle of these components cannot differ by more than a few degrees (<5°) if both are to be highly boosted. In the CR model this effectively rules out the possibility of a significant change in direction between the motion in the acceleration zone and the motion further out. It also needs to be kept in mind this model that the superluminal motion is seen in the portion of the jet that would have the largest viewing angle. It seems very unlikely that this core-in-jet model can be a viable one and our previous conclusions [26] that the radio core is un-boosted and centered on the accretion disc remains much more likely.

In fact Homan et al. [30] have difficulty explaining the brightness they see in some of the features in the mas jet of PKS 1510-089 by Doppler boosting, arguing that the brightness must be dominated by shocked emission. This is very damning for the relativistic beaming model. Also, as they too admit, the high levels of fractional polarization they detect in the outer edge of the mas jet suggests that the bow shock is seen from the side, which would be the case if the viewing angle of the jet was large as is being suggested here, instead of coming towards us, as would be the case in their model.

From the above examination it is concluded here that most of the flux from the core component is un-boosted, with almost all of the boosted radiation in these sources originating then in the inner jet. This conclusion is also consistent with the fact that it is only in the inner jet that apparent superluminal motions have been conclusively detected. We are now interested in determining what percentage of the total flux would have come from the inner jet in the finding survey.

RELATIVE STRENGTHS OF THE BOOSTED AND UN-BOOSTED RADIATION

In the VLBA contour plots of the core and inner mas jets of 132 radio-loud AGN galaxies (radio galaxies, BLLacs and quasars) obtained at 15 GHz [7], the flux density from the unresolved, compact core component dominates that from the pc-scale, inner jet in most cases. Twenty of these sources, chosen mainly because they have very high β_{app} values, are included in **Table 1**. Here $\beta_{app} = v_{app}/c$, where v_{app} is the apparent linear speed in the jet obtained from the observed angular motion, assuming that the source is located at a distance determined from its redshift. Because of their high apparent superluminal motions the jets of these sources must have very small viewing angles if these motions are to be explained in the relativistically beaming model. From their contour plots it can be seen that the inner, pc-scale jets of these sources almost always cover several beam areas. Because the core component is unresolved it is assumed that a small part of the inner jet component lying at the base of the pc-scale jet would have been included in the peak flux of the core. This can represent only a very small portion of the core flux, however, when the inner jet covers several beam areas, and the entire inner jet radiation component is itself, in most cases, much smaller than the peak core component.

Thus, although the core and inner end of the jet cannot be resolved, the component of the flux coming from the inner jet but included in the core peak flux will be negligible. From this we have estimated the approximate 2 cm flux from the pc-scale jet, S_{in}, using the relation $S_{in} = S_{total} - S_{peak}$, where S_{total} and S_{peak} are flux values obtained with the VLBA and have been taken from **Table 3** of [7]. S_{in} has been included in column 7 of **Table 1**. Columns 5 and 6 give the source flux densities measured in the finding surveys at 178 MHz or 400 Mhz, taken from the Dixon catalog [31], the Parkes catalog [32], the 4C (+20° to +40°) catalog [33], and the 4C (–7° to +20° and +40°º to +80°) catalog [34]. S_{ext}, included in column 8 of **Table 1**, represents the flux from the external, kpc-scale, jet component taken from [35] and [36].

From the examination carried out in the previous section, for the purposes of this investigation we shall assume, 1) that the radiation from the core, S_{peak}, is almost entirely unboosted, 2) that the material in the inner jet is almost certainly to be moving relativistically in the CR model and will be boosted if its direction is towards us, and 3) that most of the material in the external, kpc-scale jet in these core-dominant sources is not moving relativistically and will therefore not be boosted.

In **Table 1**, column 9 lists F_{IJ}, the ratio of the inner jet flux found at 2 cm, where the resolution is adequate to resolve it, to the total flux found at the low frequencies of the finding surveys, expressed as a percentage. We assume here that the spectral index of the jet is flat even though there is a good chance that all, or at least part, of the inner jet may be located inside the low-frequency photosphere, which would prevent its detection at the low radio frequencies used in the finding surveys. If this is the case the value of F_{IJ} would be even smaller than the value listed.

To be considered dominant F_{IJ} needs to make up more than 90 percent of the total flux. As can be seen in Table 1, no source comes close to this. Even when it is assumed in column 10 that the external kpc jet flux is also boosted, the entire jet component, F_{EJ}, is also far from dominating the total source flux. It is therefore not possible for the inner jet, or even the entire jet, to have introduced into the finding survey a strong selection effect that would have preferentially chosen sources with small inclination angles. As noted above, this is because most of these sources would have been detected even without this small amount of boosted radiation from the jet, and their distribution of orientations must then be close to random. In particular, we note that the outer jet in PKS 1510-089 has been found, in the CR model, to be directed at an angle of between 12° - 24° from the line-of-sight [30]. This means that its radiation would not be significantly boosted.

Table 1. Percent of flux likely to be boosted for high-β_{app} sources.

Source	Alt.name	β_{app}	z	S_{178} (Jy)	S_{408} (Jy)	S_{in} (Jy)[1]	S_{ext} (Jy)[2]	F_{IJ} (%)[3]	F_{EJ} (%)[4]
0106 + 013	PKS	23	2.1	-	3.5	0.33	0.531	9.4	24.6
0149 + 218	PKS	18	1.32	-	1.9	0.16	0.025	8.4	9.7
0234 + 285	4C + 28.07	13	1.213	2.1	-	0.49	0.1	23	28.1
0333 + 321	4C + 32.14	24	1.263	2.2	-	0.4	0.072	18.2	21.6
0420 - 014	PKS	14	0.92	1.2	1.5	0.64	0.070	42	47
0850 + 581	4C + 58.17	13	1.32	2.9	-	0.24	-	8.3	8.3
0945 + 408	4C + 40.24	22	1.252	2.5	-	0.56	0.095	22	26
1156 + 295	4C + 29.45	8.9	0.729	2.8	-	0.34	0.196	12	19.1
1226 + 023	3C273	14	0.158	75	-	16.5	17.6	22	45
1508-055	PKS	31	1.18	-	8.9	0.19	-	2.1	2.1
1510-089	PKS	19	0.36	-	3.0	0.46	0.18	15	21.3
1606 + 106	4C + 10.45	30	1.226	2.7	4.4	0.33	0.26	12	22
1633 + 382	4C + 38.41	11.5	1.807	2.2	-	0.67	0.032	30	32
1641 + 399	3C345	17	0.594	10	-	3.95	1.48	39	54
1642 + 690	4C + 69.21	16	0.751	2.5	-	0.27	0.33	10	24
1730-130	PKS	23	0.90	-	6.3	1.09	0.517	17	25.5
1823 + 568	4C + 56.27	3.4	0.663	2.4	-	0.26	0.137	10.8	16
1828 + 487	3C380	15	0.692	57	-	1.0	5.43	1.7	11
2201 + 315	4C + 31.63	6.3	0.298	3.5	-	0.82	0.378	23	34
2223 - 052	3C446	32	1.404	17.3	-	0.64	0.92	4	9

[1]Flux in the Inner Jet at 2 cm where $S_{in} = S_{total} - S_{peak}$ from [7]. [2]Flux in external (kpc) jet from [35,36]. [3]F_{IJ} = Percentage of Inner Jet Flux (S_{in}) compared to total Flux in Finding Survey. Assumes a flat jet spectral index. [4]F_{EJ} = Percentage of Entire (inner and outer) jet flux compared to total flux in finding survey. Assumes a flat jet spectral index.

Since the sources involved here are radio-loud AGNs found in early surveys made at low frequencies and with large beamwidths, in most cases their detection will have been based on the total flux. Here we find that instead of the boosted radiation representing at least 90% of the total flux, it is the unboosted radiation that is dominant, appearing to represent ~90% of the total flux density from many of the radio loud quasars with high β_{app} values. This situation will only be worsened if the spectral index of the inner jet is not flat, as assumed here, and actually falls off at the low frequencies of the finding surveys.

It is worth noting that this model, where the jets turn on and off and are closer to the plane of the sky, would then easily explain why [30] found no evidence for a counter-jet in PKS 1510 - 089 by simply interpreting the arcsec jet, lying ~180° from the milliarcsec jet, as the counter-jet. This would require, as is being proposed here for most of these sources, that the jet and counter-jet are both at large viewing angles instead of being viewed endon as proposed by [30]. In this scenario there is also evidence that the jets in PKS 1510 - 089 switch on and off, as is required to explain intrinsic one-sidedness. Furthermore, as noted above, the polarization detected at the end of the mas jet is also consistent with this scenario. Although in the CR model the superluminal motion of the blobs in the pc-scale jet of 3C279 requires a viewing angle of i = 2° to explain [37], our results indicate that the viewing angle of 30° to 40° found for the inner jet by [38] may actually be the correct one.

In summary, when the strengths of the boosted and unboosted radiation are compared, only a very small percentage of the total flux density of these radio loud quasars can be coming from material that is being ejected in a tightly confined beam and at relativistic speeds, and it must be concluded that Doppler boosting is unlikely to have played a significant role in the finding surveys in which radio-loud quasars were detected. Astronomers have been aware of this problem ever since the relativistic beaming model was first proposed to explain superluminal motion. At that time there were some concerns that it might be difficult to explain the large number of jets with small viewing angles that seemed to be required, and it was partly this concern that Scheuer [39] was expressing when he stated that "it is the theoretician's duty to look for ways of escape if the observations should confound the predictions." For a review see [3].

DISCUSSION

If there are no selection effects operating the distribution of orientations for these sources will be given by the well-known sini curve in **Figure 2** (represented by the solid curve), which is the curve predicted for a random distribution of viewing angles. In **Figure 2** the vertical axis represents the number of sources expected in each 2 degree-wide inclination bin, for inclinations between 0° and 90°, assuming the finding survey found a total of 500 sources. By summing sources at 2, 4 and 6 degrees it is found that only 6 (1%) of the detected sources would have had inclination angles that are close to the line of sight (below 8°). Also included in **Figure 2** (circular points) are β_{app} values calculated for some of the sources studied by [7,8]. These are plotted vs. viewing angle on the same scale. In this case the jet viewing angles are those required in the CR model to explain the measured $ß_{app}$ values, as calculated by Hovatta et al. [37,40] for 67 jetted sources. In this plot 56 of the 67 sources, or 84%, require jet viewing angles i< 8°. It also shows that if the redshifts are cosmological almost all sources with $ß_{app}$ > 5 are required to have viewing angles i< 8°, whereas we have just shown that almost none can fall in this category because only a very small percentage of the flux can be boosted resulting in a close-to-random distribution. It is also worth noting that even if the predicted number distribution curve were flat, or even cosi, it would be impossible to reconcile it with the observed number distribution obtained using $ß_{app}$ in **Figure 2**.

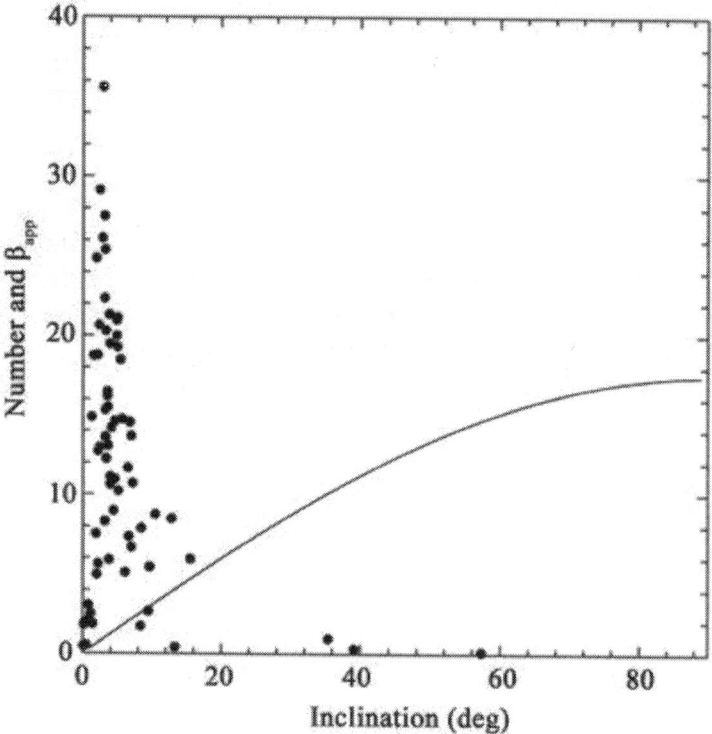

Figure 2. (Solid line) The sini number vs. jet viewing angle distribution expected if no selection effects are active in source finding surveys and 500 sources are found. (points) Actual $ß_{app}$ vs. jet viewing angle distribution from [37], assuming quasar redshifts are cosmological.

Although it has been assumed in **Figure 2** that 500 sources above 1 Jy would be found, this number may be too high for radio-loud quasars. The Wall and Peacock sample [41] contains 233 bright extragalactic radio sources found in the major centimeter-wavelength surveys at Parkes, Green Bank, and Bonn, and is complete to 2 Jy at 2.7 GHz. The list of bright radio sources at 178 MHz [42] contains 181 sources. The Kühr sample [43,44] contains 518 sources and is complete to 1 Jy at 5 GHz. The 3CR sample has a similar number.

However, each of these samples contains many mature radio galaxies that are not part of the high-redshift quasar sample considered here. For example, the Kühr sample contains 165 radio galaxies (see figure 1 of [5]). The nature of the redshifts of these mature radio galaxies is not in question and is assumed to be cosmological. The complete radio-loud quasar sample (i.e. quasars and BL Lacs) can therefore be assumed to contain closer to 330 sources, which is considerably less than 500.

The 117 radio-loud quasars and BL Lacs with jets included in the Kellermann [7,8] sample thus represent many of those found mainly in the early surveys, which would have been found because of their strong, total flux density, almost all of which is un-boosted. Therefore, for a random distribution, less than ~3 of these sources would be expected to have jet viewing angles less than 8°. In that study 86% of the sources had $\beta_{app} > 1$, 63% had $\beta_{app} > 3$, 50% had $\beta_{app} > 5$. There were 16 3C-sources in their sample and 50% of these had $\beta_{app} > 3$. There were 21 4C-sources and, of these, 80% had $\beta_{app} > 3$ and 65% had $\beta_{app} > 5$. Overall, there were 34 sources, or 26%, with $\beta_{app} > 10$ (i< 8°). Of these, 25 are PKS or 4C sources, or both. Almost all of these (23) are high redshift sources, with redshifts greater than z = 0.6.

In the Kühr sample, approximately 75% of the 269 quasars with measured redshifts (~200 sources) and measured spectral index, have reasonably flat spectra. If 26% of these have $\beta_{app} > 10$, like those in the Kellermann sample, [8] then ~50 of these would have to have viewing angles less than 8°, where less than ~3 are expected for a random distribution. The Kellermann sample was drawn from the list of radio-loud sources found in the original surveys and if these lists contain only a very few sources with small viewing angles, no matter how the sources are later chosen it cannot change the total number with small viewing angles that are available to be chosen. Since the evidence then indicates that almost none of the sources with high-β_{app} values can have been preferentially selected because of Doppler boosting, almost all must have viewing angles i> 8°.

The results found here also mean that the term blazar needs to be more clearly defined. This term has come to represent a quasar or BL Lac object whose variability results mainly from the fact that the jet is pointed in our direction [36]. It now can imply only that the flux density fluctuations seen in AGNs are due simply to the fact that the central engine is currently swallowing, and spitting out, new in-falling material, without any implication that the jet is pointed in our direction. This explanation also fits the observations better since the fluctuations in 3C279 [26,27] and other radio variable sources are observed to be associated mostly with the growth and decline of the number of blobs moving away from the core at any given time, and not with simultaneous fluctuations in all blobs. The latter might be expected if, as has been previously suggested, the fluctuations were due to small changes in the viewing angle of jets closely aligned with the line-ofsight. The fact that this is not seen also agrees with our finding that the jet viewing angles are large in almost every case.

How Complete Is the Radio-Loud AGN Sample?
If, for example, we assume that 95% of the radio radiation from radio-loud quasars is coming from material that is moving out in a jet at relativistic speeds, then because of the relativistic beaming effect where the radiation is enhanced in the direction of motion, those sources with their jets pointed in our direction would be significantly stronger than those whose jets have large viewing angles. In detection-limited finding surveys many of the sources whose jets have large viewing angles would then fall below the detection limit while those pointed in our direction would be detected. In this scenario the high percentage of radio-loud quasars requiring small viewing angles could be explained as representing that few percent of sources in a random sample that have small viewing angles, while the remaining ~95% of the sample lies below the detection limit. However, if, on the other hand, only a small percentage of the source radiation comes from material that is moving out at relativistic

speeds, the flux from those sources with large jet viewing angles would not differ significantly from those with small viewing angles. In this scenario almost all of the radio-loud sources would be detected and the sample would be essentially complete. We have shown above that it is this latter situation that is most likely to be the correct one.

In **Figure 3**, the number of sources from **Table 1** is plotted versus F_{IJ}, the relative percentage of boosted to unboosted flux. The number peaks near $F_{IJ} = 20\%$. If the boosted inner jet is only 20% of the total flux, the strengths of radio-loud AGNs with larger viewing angles would not be expected to be significantly fainter than those with small viewing angles even if the entire inner jet disappeared. The radio-loud sample would then be expected to be reasonably complete above 1 Jy for all viewing angles.

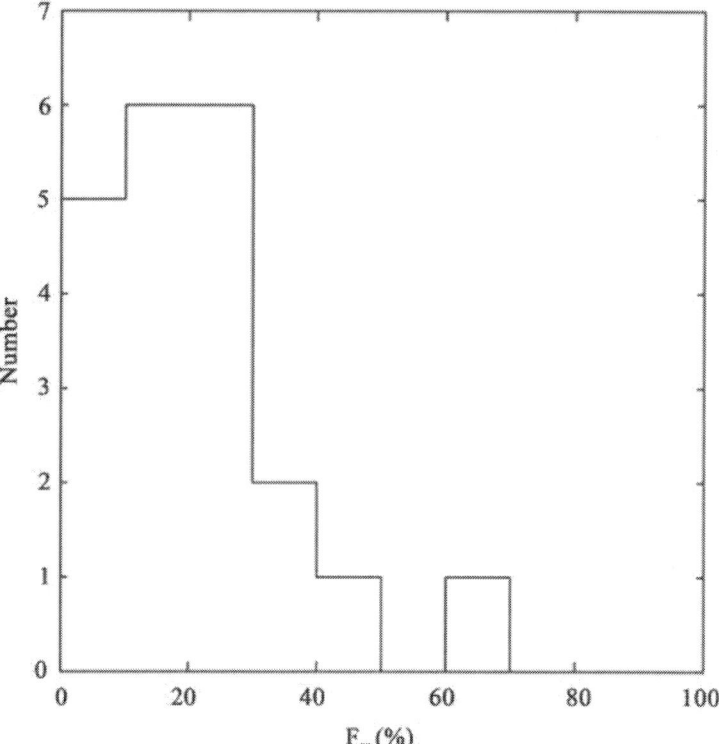

Figure 3. Number of sources from **Table 1** plotted versus F_{IJ}, the percentage of the flux in the inner jet compared to the total flux in the finding survey.

It is concluded that the 330 radio-loud quasars in the Kühr sample make up close to a complete radio-loud sample. This makes the high percentage of sources observed with large apparent superluminal motions in the CR model very difficult to explain statistically.

In summary, when the Doppler boosted component is small compared to the total source flux, as found here, it can be concluded that the sample of radio-loud sources detected in a finding survey will represent almost all of the radio-loud sources and only a few percent of them can have small viewing angles. In this situation some explanation other than relativistic beaming must be found to explain the high percentage of sources exhibiting large apparent superluminal motions.

Other Radio Selection Effects

There may still be some radio selection effects present that are related to viewing angle but unrelated to Doppler boosting. For example if there is a torus surrounding the central object it can block some of the radiation coming from the central compact object when inclinations are large (near edge-on). Recently Lovegrove et al. [45-46] have measured the opening angles and inclinations for 55 radio quiet quasars. They found opening angles near 78° in these objects and the distribution of inclinations from [45] has been plotted in **Figure 4**, where it can be seen that for small inclination angles the number distribution follows the sini curve closely. The vertical dashed line indicates the inclination angle (78°/2) above which the torus prevents the central compact object from being viewed. In fact, when viewed in this manner, the results in **Figure 4** suggest that the actual opening angle may be slightly smaller (60° - 65°) than the 78° reported by [45]. For large inclination angles the torus has clearly affected the detection of these radio-quiet objects. However, it is obvious from **Figure 4** that the radio-quiet distribution does follow a sini curve for viewing angles that are unaffected by the opening angle cut-off.

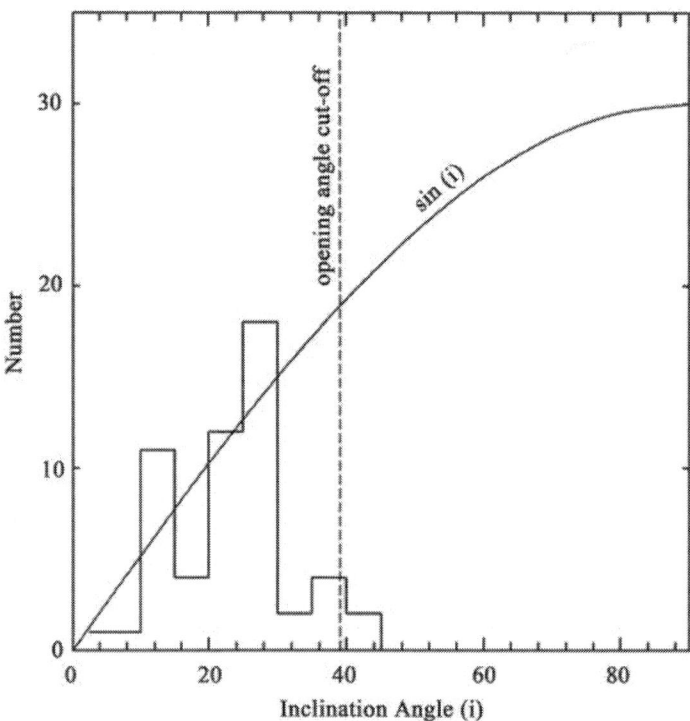

Figure 4. Viewing angle distribution for 55 radio-quiet quasars from [45]. The vertical dashed line indicates the viewing angle above which the opening angle of 78° found by these authors would be expected to affect the number distribution.

Radio Quiet Objects as the Parent Population of Radio Loud Objects

There is clear confirmation from **Figure 4** that without a dominant Doppler boosted component present few sources with viewing angles less than 10° will be detected. If the radio-quiet quasars really represent the parent sample from which the radio-loud quasars are drawn, when the radioloud quasars contain only a small boosted component, as found here, the radio-loud sources must then be radio loud because they are closer, but they are still expected to have the same sini distribution of orientations as given by the solid curve in **Figure 4**. The radio-loud distribution given by the filled circles in **Figure 2**, and determined assuming that the distance to these objects is reliably given by their redshifts, is clearly incompatible with the sini distribution found for the radio-quiet sources in **Figure 4**. This is strong confirmation that if the Doppler boosted component is small, as found here, the redshifts of the objects in **Figure 2** cannot be an accurate indication of distance.

CONCLUSION

It is concluded here that Doppler boosting could not have played a significant role in finding radio-loud, high redshift quasars because the component of their radiation that comes from material being ejected outwards at relativistic speeds, and in a tightly confined jet, is insignificant compared to the total flux obtained in the low-frequency finding surveys. This is true even if the radiation from the kpc-scale jet is from material that is moving relativistically, and is especially true for sources with large radio lobes. Without a highly directed, relativistically beamed component that dominates the source flux density, sources cannot be preferentially selected with small jet viewing angles and the resulting distribution of jet viewing angles will then be close to that of a random one (sini). In this case almost all will have viewing angles much greater than 8° and even the flux from the inner jet will be unboosted. This means that relativistic ejections with small jet viewing angles cannot be used to explain the observed superluminal motions seen in high-redshift quasars. Although this problem can easily be resolved by bringing these sources closer and accepting intrinsic redshift components in high redshift quasars, this solution has so far been found unacceptable by most astronomers. At the very least, a new explanation for superluminal motion that does not involve relativistic beaming will need to be found if the redshifts of high redshift quasars are to remain cosmological.

ACKNOWLEDGEMENTS

I thank S. Comeau and D. McDiarmid for helpful comments when this manuscript was being prepared.

REFERENCES

1. M. López-Corredoira, "Pending Problems in QSOs," International Journal of Astronomy and Astrophysics, Vol. 1, No. 2, 2011, pp. 73-82. doi:10.4236/ijaa.20011.12011

2. M. J. Rees, "Appearance of Relativistically Expanding Radio Sources," Nature, Vol. 211, 1966, pp. 468-470. doi:10.1038/211468a0

3. J. A. Zensus and T. J. Pearson, "Superluminal Radio Sources: Introduction," In: J. A. Zensus and T. J. Pearson, Eds., Superluminal Radio Sources, Cambridge University Press, Cambridge, 1987

4. J. V. Narlikar, "Noncosmological Redshifts," Space Science Reviews, Vol. 50, No. 3-4, 1989, pp. 523-614. doi:10.1007/BF00228382

5. M. B. Bell, "Further Evidence That the Redshifts of AGN Galaxies May Contain Intrinsic Components," Astrophysical Journal Letters, Vol. 667, No. 2, 2007, pp. L129- L132.

6. K. R. Lind and R. D. Blandford, "Semidynamical Models of Radio Jets: Relativisitc Beaming and Source Counts," The Astrophysical Journal, Vol. 295, 1985, pp. 358-367.doi:10.1086/163380

7. K. I. Kellermann, R. C. Vermeulen, J. A. Zensus and M. H. Cohen, "Sub-Milliarcsec Imaging of Quasars and Active Galactic Nuclei," The Astronomical Journal, Vol. 115, No. 4, 1998, pp. 1295-1318. doi:10.1086/300308

8. K. I. Kellermann, et al., "Sub-Milliarcsec Imaging of Quasars and Active Galactic Nuclei. III. Kinematics of Parsec-Scale Radio Jets," The Astrophysical Journal, Vol. 609, No. 2, 2004, pp. 539-563. doi:10.1086/421289

9. C. M. Urry and P. Padovani, "Unified Schemes for RadioLoud Active Galactic Nuclei," Publications of the Astronomical Society of the Pacific, Vol. 107, No. 715, 1995, pp. 803-845. doi:10.1086/133630

10. Y. Y. Kovalev, M. L. Lister, D. C. Homan and K. I. Kellermann, "The Inner Jet of the Radio Galaxy M87," Astrophysical Journal Letters, Vol. 668, No. 1, 2007, pp. L27-L30.

11. J. A. Biretta, W. B. Sparks and F. Macchetto, "Hubble Space Telescope Observations of Superluminal Motion in the M87 Jet," The Astrophysical Journal, Vol. 520, No. 2, 1999, pp. 621-626. doi:10.1086/307499

12. M. B. Bell, "Evidence in Support of the Local Quasar Model from Inner Jet Structure and Angular Motions in Radio Loud AGN," 2007, arXiv:0711.4531

13. J. G. Bolton, F. F. Gardner and M. B. Mackey, "The Parkes Catalogue of Radio Sources, Declination Zone −20° to −60°," Australian Journal of Physics, Vol. 17, No. 3, 1964, pp. 340-372. doi:10.1071/PH640340

14. G. K. Miley, "The Radio Structure of Quasars—A Statistical Investigation," Monthly Notices of the Royal Astronomical Society, Vol. 152, 1971, pp. 477-490.

15. P. Alexander and J. P. Leahy, "Ageing and Speeds in a Representative Sample of 21 Classical Double Radio Sources," Monthly Notices of the Royal Astronomical Society, Vol. 225, 1987, pp. 1-26.

16. K. Cleary, C. R. Lawrence, J. A. Marshall, L. Hao and D. Meier, "Spitzer Observations of 3C Quasars and Radio Galaxies: Mid Infrared Properties of Powerful Radio Sources," The Astrophysical Journal, Vol. 660, No. 1, 2007, pp. 117-145. doi:10.1086/511969

17. C. P. O'Dea, et al., "Physical Properties of Very Powerful FRII Radio Galaxies," Astronomy and Astrophysics, Vol. 494, No. 2, 2009, pp. 471-488. doi:10.1051/0004-6361:200809416

18. R. A. Laing, P. Parma, H. R. De Ruiter and R. Fanti, "Asymmetries in the Jets of Weak Radio Galaxies," Monthly Notices of the Royal Astronomical Society, Vol. 306, No. 3, 1999, pp. 513-530. doi:10.1046/j.1365-8711.1999.02548.x

19. R. A. Laing and A. H. Bridle, "Relativistic Models and the Jet Velocity Field in the Radio Galaxy 3C 31," Monthly Notices of the Royal Astronomical Society, Vol. 336, No. 1, 2002, pp. 328-352. doi:10.1046/j.1365-8711.2002.05756.x

20. R. A. Laing and A. H. Bridle, "Dynamical Models for Jet Deceleration in the Radio Galaxy 3C 31," Monthly Notices of the Royal Astronomical Society, Vol. 336, No. 4, 2002, pp. 1161-1180. doi:10.1046/j.1365-8711.2002.05873.x

21. R. A. Laing, J. R. Canvin, A. H. Bridle and M. J. Hardcastle, "A Relativistic Model of the Radio Jets in 3C 296," Monthly Notices of the Royal Astronomical Society, Vol. 372, No. 2, 2006, pp. 510-536. doi:10.1111/j.1365-2966.2006.10903.x

22. R. A. Laing and A. H. Bridle, "Jet-Environment Interactions in FRI Radio Galaxies," In: T. A. Rector and D. S. Young, Eds., Extragalactic Jets: Theory and Observation from Radio to Gamma Rays, ASP Conference Series, Vol. 386, 2008, pp. 70-79.

23. A. P. Marscher, "Effects of Nonuniform Structure on the Derived Physical Parameters of Compact Synchrotron Sources," The Astrophysical Journal, Vol. 216, 1977, pp. 244-256.doi:10.1086/155467

24. A. P. Marscher, "The Core of a Blazar Jet," In: T. A. Rector and D. S. Young, Eds., Extragalactic Jets: Theory and Observations from Radio to Gamma Rays, ASP Conference Series, Vol. 386, 2008, pp. 437-443.

25. A. P. Marscher, "Jets in Active Galactic Nuclei," 2009, arXiv:0909.2576v1[astroph.HE]

26. M. B. Bell and S. P. Comeau, "The Point of Origin of the Radio Radiation from the Unresolved Cores of RadioLoud Quasars," Astrophysics and Space Science, Vol. 325, No. 1, 2010, pp. 31-36. doi:10.1007/s10509-009-0162-z

27. R. Chatterjee, et al., "Correlated Multi-Wave Variability in the Blazar 3C 279 from 1996 to 2007," The Astrophysical Journal, Vol. 689, No. 1, 2008, pp. 79-94. doi:10.1086/592598

28. S. G. Jorstad, et al., "Flaring Behavior of the Quasar 3C 454.3 across the Electromagnetic Spectrum," 2010, arXiv:1003.4293 [astroph.CO]

29. M. Sikora, R. Moderski and G. M. Madejski, "3C 454.3 Reveals the Structure and Physics of Its 'Blazar Zone'," The Astrophysical Journal, Vol. 675, No. 1, 2008, pp. 71-78.doi:10.1086/526419

30. D. C. Homan, J. F. C. Wardle, C. C. Cheung, D. H. Roberts and J. M. Attridge, "PKS 1510-089: A Head-on View of a Relativistic Jet," The Astrophysical Journal, Vol. 580, No. 2, 2002, pp. 742-748. doi:10.1086/343894

31. R. S. Dixon, "A Master List of Radio Sources," Astrophysical Journal Supplement, Vol. 20, 1970, pp. 1-503. doi:10.1086/190216

32. J. Ekers, "The Parkes Catalogue of Radio Sources: Declination Zone +20° to −90°," Australian Journal of Physics Astrophysical Supplement, Vol. 7, 1969, pp. 3-75.

33. J. Pilkington and P. Scott, "A Survey of Radio Sources Between Declinations 20° and 40°," Memoirs of the Royal Astronomical Society, Vol. 69, 1964, pp. 183-192.

34. J. Gower, P. Scott and D. Wills, "A Survey of Radio Sources in the Declination Ranges −07° to 20º and 40° to 80°," Memoirs of the Royal Astronomical Society, Vol. 71, 1967, pp. 49-144.

35. D. Murphy, I. Browne and R. Perley, "VLA Observations of a Complete Sample of Core-Dominated Radio Sources," Monthly Notices of the Royal Astronomical Society, Vol. 264, No. 2, 1993, pp. 298-318.

36. P. Kharb, M. L. Lister and N. J. Cooper, "Extended Radio Emission in MOJAVE Blazars: Challenges to Unification," The Astrophysical Journal, Vol. 710, 2010, pp. 746-782.

37. T. Hovatta, E. Valtaoja, M. Tornikoski and A. Lahteenmaki, "Doppler Factors, Lorentz Factors and Viewing Angles for Quasars, BL Lacertae Objects and Radio Galaxies," Astronomy and Astrophysics, Vol. 494, No. 2, 2009, pp. 527-537. doi:10.1051/0004-6361:200811150

38. E. A. Carrara, Z. Abraham, S. C. Unwin and J. A. Zensus, "The Milliarcsecond Structure of the Quasar 3C 279," Astronomy and Astrophysics, Vol. 279, No. 1, 1993, pp. 83-89.

39. P. Scheuer, "Tests of Beaming Models," In: J. A. Zensus, and T. J. Pearson, Eds., Superluminal Radio Sources, Cambridge University Press, Cambridge, 1987, pp. 104-113.

40. T. Hovatta, E. Valtaoja, M. Tornikoski and A. Lahteenmaki, "Doppler Factors, Lorentz Factors, and Viewing Angles for Quasars, BL Lacertae Objects and Radio Galaxies (Erratum)," Astronomy and Astrophysics, Vol. 498, No. 3, 2009, pp. 723-723.doi:10.1051/0004-6361:200811150e

41. J. V. Wall and J. A. Peacock, "Bright Extragalactic Radio Sources at 2.7 GHz, III. The All-Sky Catalogue," Monthly Notices of the Royal Astronomical Society, Vol. 216, No. 1, 1985, pp. 173-192.

42. R. A. Laing, J. M. Riley and M. S. Longair, "Bright Radio Sources at 178 MHz: Flux Densities, Optical Identifications and the Cosmological Evolution of Powerful Radio Galaxies," Monthly Notices of the Royal Astronomical Society, Vol. 204, 1983, pp. 151-187.

43. H. Kühr, A. Witzel, I. I. K. Pauliny-Toth and U. Nauber, "A Catalogue of Extragalactic Radio Sources Having Flux Densities Greater than 1 Jy at 5 GHz," Astronomy and Astrophysics Supplement Series, Vol. 45, 1981, pp. 367-430.

44. M. Stickel, K. Meisenheimer and H. Kühr, "The Optical Identification Status of the 1 Jy Radio Source Catalogue," Astronomy and Astrophysics Supplement Series, Vol. 105, No. 2, 1994, pp. 211-234.

45. J. Lovegrove, R. E. Schild and D. Leiter, "Discovery of Universal Elliptical Outflow Structures in Radio-Quiet Quasars," 2010, arxiv:1003.5497.

46. J. Lovegrove, R. E. Schild and D. Leiter, "Discovery of Universal Outflow Structures above and below the Accretion Disc Plane in Radio-Quiet Quasars," Monthly Notices of the Royal Astronomical Society, Vol. 412, No. 4, 2011, pp. 2631-2640. doi:10.1111/j.1365-2966.2010.18082.x

CITATION

M. Bell, "Doppler Boosting May Have Played No Significant Role in the Finding Surveys of Radio-Loud Quasars,"*International Journal of Astronomy and Astrophysics*, Vol. 2 No. 1, 2012, pp. 52-61. doi:10.4236/ijaa.2012.21008.

Chapter 4

Total Solar Flux Intensity at 11.2 GHz as an Indicator of Solar Activity and Cyclicity

JuhaKallunki, MinttuUunila

Metsähovi Radio Observatory, Aalto University, Kylmälä, Finland

ABSTRACT

In this paper we present an overview of solar radio observations at 11.2 GHz on Metsähovi Radio Observatory (MRO). The data were observed during the solar cycles 23 and 24 (2001-2013) both in solar maxima and minimum. In total, 180 solar radio bursts, with varying intensities and properties, were observed. We compare our data series with other similar data sets. A good correlation can be found between the data series. It is concluded that one can conduct scientifically significant solar radio observations with a low cost instrument as the one presented in this paper.

INTRODUCTION

Metsähovi RT-1.8, located at Metsähovi Radio Observatory (MRO), Aalto University (Helsinki Region, Finland, GPS: N60°13.04!, E24°23.35!), is a radio telescope with a 1.8 m dish diameter dedicated for continuous solar observations. The telescope has a beam size of 81.6 arc min and its system noise temperature is 270 K. It observes the total radiation of the Sun at a frequency of 11.2 GHz. The emission measured at 11.2 GHz originates from lower corona. The Quiet Sun Level (QSL) is around 12,000 K at 11.2 GHz. The radio telescope is used for observing solar radio bursts, as it acts as a detector of general solar activity. It also studies on solar oscillations have been done. High sampling rate (5 kHz) enables studying flare fine structure, including short periodic oscillation phenomena. The radio telescope has no protective radome, therefore it is vulnerable to prevailing weather conditions. Around 200 solar radio bursts have been detected since their launch in 2001. Full documentation of the Metsähovi RT-1.8 can be found in [1] . The telescope has a logarithmic output which can be utilized in the case of strong bursts in which case the linear output that is used in this study, would saturate. The logarithmic output was not enabled during this research. The changing radio emission indicates variation of solar magnetic activity. The birth of radio emission is affected by all plasma parameters, for example, temperature and density. This signifies the importance of radio observations and introduces interesting information on the subject. Gyromotion of thermal electrons in a presence of a magnetic field causes thermal gyroresonance emission. In active regions the magnetic field may have the strength to render the corona optically thick to enable gyroresonance absorption at the frequency range of 1 - 18 GHz.

The first observation of a solar radio burst was measured in the 1940's. Thus, there are only few observations and their time span is only about 70 years. Data from Metsähovi RT-1.8 have been used earlier, for example, by [2] who studied solar oscillations during a solar radio burst (flare). They found various oscillation periods between 1.9 and 12.8 minutes. They assumed that some of these periods corresponded to the signatures of largescale transverse oscillations of coronal loops. This study presents a new long-term time series of solar radio burst data which is compared to several existing time series. At MRO a similar time series has been produced at a frequency of 37 GHz [3] .

CLASSIFICATION OF SOLAR RADIO BURSTS

In this study solar radio bursts were categorized in three different classes from I to III (I being the weakest and III being the strongest) on the basis of their strengths. Due to the fact that the data is not perfectly calibrated, we divided it into the three classes. In Figure 1 one can see an example of a typical measurement day (13.5.2013) with one Class III burst. Observations cover a time period from 1/2001 to 12/2013 excluding February, March and October each year due to satellite interference. Furthermore, some other months were excluded due to technical issues. A monthly average number for the bursts is calculated. In Figure 1 Quiet Sun Level (QSL) is the median of the data set (sunshine time) which in this case is a rough estimate. Its SFU can be calculated with Equation (1) in which f is the frequency, 11.2 GHz [4] :

$$SFU = 2.79 \times 10^{-5} \times f^{1.748}, \ 6000 - 400,000 \ \text{MHz.} \tag{1}$$

For measurement displayed in Figure 1, the result is 450 SFU. If a burst exceeds 500 SFU, the result will saturate. By choosing Class I to be as high as 1.2× QSL, it is ensured that no artifacts are introduced into the results. At MRO long time series of solar observations are possible only during the summers due to the Northern location of the site. The biggest reason that causes problems is the weather, for example, wind and rain. As an example of such a weather effect can be seen in Figure 1, a dip on intensity curve at around 15:00. Furthermore, during winter the Sun is below the horizon for the most part of the day. From the beginning of 2014 a new calibration method, that exploits the use of a calibration noise diode, has been in use allowing a more accurate division of the solar radio bursts. Calibration is now performed daily.

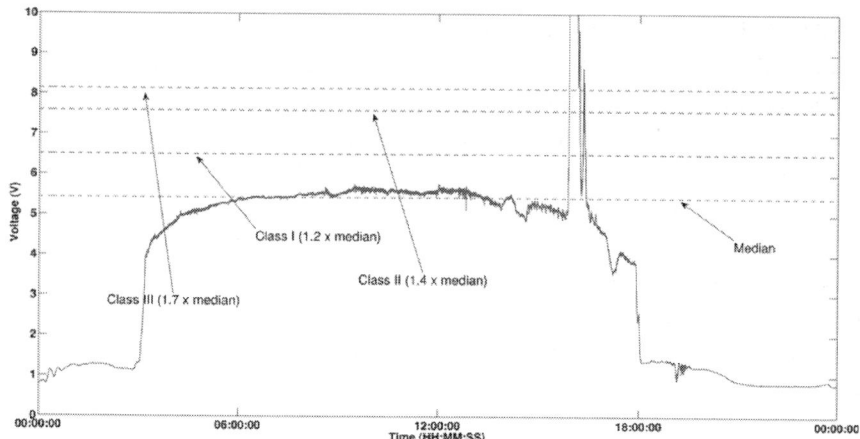

Figure 1. Example of a 24-hour measurement on 13.5.2013.

RESULTS

A summary of solar radio bursts is listed in Table1 It is surprising that the number of events in Class II and III is about the same.

In Figure 2 and Table 2 a distribution of the solar radio bursts into different classes during 2001-2013 can be seen. Monthly averages of total number events are listed in Table3 Near the beginning and end of the data collection period the amount of Class III and II bursts are at their highest. Note, that in the middle of the data collection period no solar radio bursts were recorded (Table 3) or their strength was too low to be recorded. This also indicates a very quiet solar minimum. It can also be seen from Table 3 that solar cycle 24 is weaker than cycle 23 as the number of bursts is lower.

[5] has recorded a similar time series as in Figure 2, but the data is not comparable because it is from an earlier time period. Also in [6] and [7] similar solar radio burst time series have been introduced but they have too been measured at different frequencies and at a previous period in time and, thus, are not fully comparable.

When the data was compared to total solar radio flux at 2.8 GHz measured at Dominion Radio Astrophysical Observatory in Canada [8] , it was noticed that the solar events measured at Metsähovi at 11.2 GHz appear to correlate with the Dominion total solar radio flux curve quite accurately (Figure 3). It should also be noted that the observations at 2.8 GHz correlate accurately with sunspot numbers.

Due to the architecture of the telescope, for example, the lack of a radome, and thus the influence of changing weather to the results, a plausible total solar radio flux curve cannot be created and directly compared with Dominion Radio Astrophysical Observatory's total solar radio flux (Figure 3).

When the MRO data was compared to data from two of the telescopes belonging to Radio Solar Telescope Network [9] Lear month, Australia and San Vito, Italy [10] , it could be concluded that again the data correlated even though the locations of the telescopes are rather far away from each other, which in turn explains why the number of the events at different locations differs (Figure 4). The size of the telescopes is about the same. Please also note that the instruments and their sensitivities differ from each other, which further explains the variation in the number of the solar events recorded at different observatories. Yet another explanation to the variation is the high QSL limit that was chosen for the MRO data. Details about the different observatories are listed inTable4

CONCLUSIONS

It can be concluded that the solar radio burst data observed with MRO's RT-1.8 telescope at 11.2 GHz are comparable to the data observed at 2.8 GHz at Dominion, Canada, 15.4 GHz at Learmonth, Australia and 8.8 GHz at San Vito, Italy. About one half of the data recorded during 13 years belong to our Class I and the other half to Classes II and III, both of which have about the same number of events. In the future, similar analysis should also be performed to Lear month and San Vito data. It is rather remarkable that a low cost instrument that utilizes commercial components such as ones used in television satellite technique, like MRO's RT-1.8, enables vast and scientifically significant observations. The instrument can be used for both cyclicity and single event analysis.

However, only one 11-year period has been measured so far. In the future MRO will continue to prolong the time series even further to allow investigating solar cyclicity thoroughly.

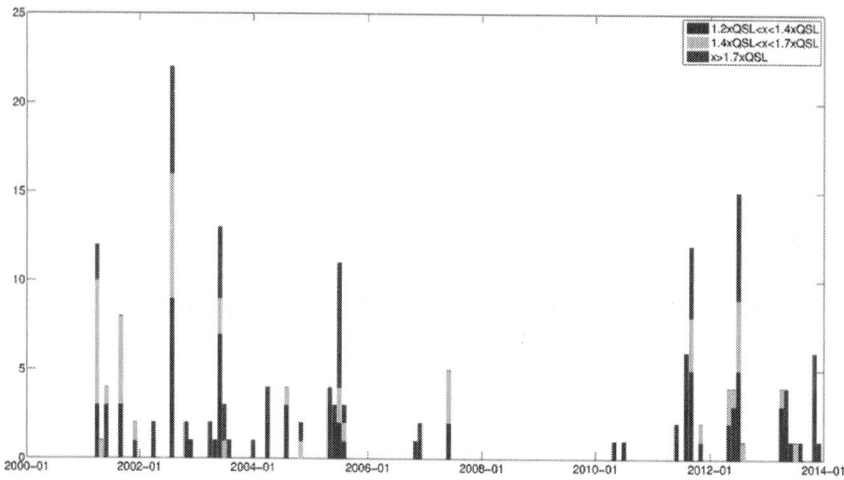

Figure 2. Classification of solar flares on basis of their intensity.

Table 1. Classification of solar radio bursts

Class	X QSL	Number of events	Percentage value (%)
I	1.2 - 1.4	84	46.7
II	1.4 - 1.7	47	26.1
III	>1.7	49	27.2
Total	-	180	100.0

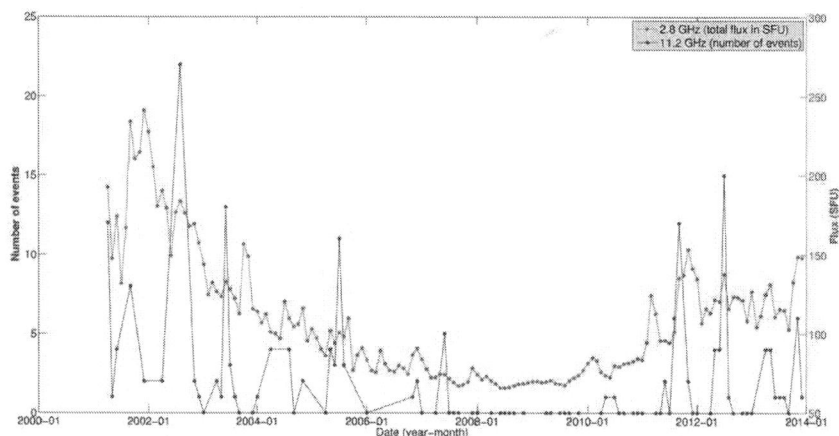

Figure 3. Comparison between number of the solar radio event at 11.2 GHz (blue) and total solar radio flux at 2.8 GHz from Dominion Radio Astrophysical Observatory, Canada (green).

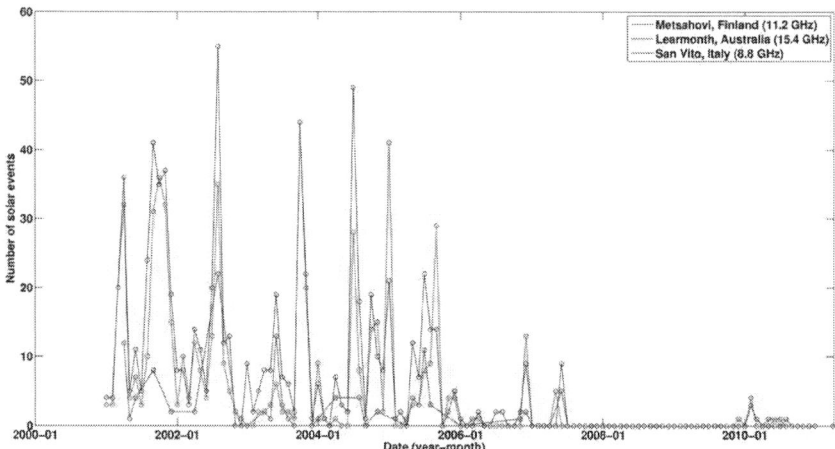

Figure 4. Comparison of number solar flares between Metsähovi (blue), Learmonth (red) and San Vito (black) between 2001 and 2010.

Table 2. Solar radio bursts 2001-2013 per month.

Month	Classes			
	Class I	Class II	Class III	Total number of events
20010401	2	7	3	12
20010501	0	1	0	1
20010601	0	1	3	4
20010901	0	5	3	8
20011201	0	1	1	2
20020401	1	0	1	2
20020801	6	7	9	22
20021101	1	0	1	2
20021201	0	0	1	1
20030101	0	0	0	0
20030401	1	0	1	2
20030501	0	0	1	1
20030601	4	2	7	13
20030701	2	1	0	3
20030801	0	0	1	1
20030901	0	0	0	0
20031201	0	0	0	0
20040101	1	0	0	1
20040401	2	0	2	4
20040801	0	1	3	4
20040901	0	0	0	0
20041101	1	1	0	2
20050401	0	0	0	0
20050501	1	0	3	4
20050601	1	0	2	3
20050701	7	2	2	11
20050801	1	1	1	3
20060101	0	0	0	0
20061101	0	0	1	1
20061201	1	0	1	2
20070101	0	0	0	0
20070401	0	0	0	0
20070601	0	3	2	5
20070701	0	0	0	0
20070801	0	0	0	0
20070901	0	0	0	0
20071201	0	0	0	0
20080101	0	0	0	0
20080401	0	0	0	0
20080601	0	0	0	0
20080701	0	0	0	0
20080801	0	0	0	0

20080901	0	0	0	0
20081101	0	0	0	0
20090401	0	0	0	0
20090501	0	0	0	0
20090701	0	0	0	0
20090801	0	0	0	0
20090901	0	0	0	0
20091101	0	0	0	0
20100401	0	0	0	0
20100501	0	0	1	1
20100701	0	0	1	1
20100801	0	0	0	0
20100901	0	0	0	0
20101101	0	0	0	0
20101201	0	0	0	0
20110101	0	0	0	0
20110401	0	0	0	0
20110501	0	0	0	0
20110601	1	0	1	2
20110701	0	0	0	0
20110801	2	0	4	6
20110901	4	3	5	12
20111101	0	1	1	2
20111201	0	0	0	0
20120101	0	0	0	0
20120401	0	0	0	0
20120501	0	2	2	4
20120601	0	1	3	4
20120701	6	4	5	15
20120801	0	1	0	1
20120901	0	0	0	0
20121101	0	0	0	0
20121201	0	0	0	0
20130101	0	0	0	0
20130401	0	1	3	4
20130501	2	0	2	4
20130601	0	0	1	1
20130701	0	1	0	1
20130801	0	0	1	1
20130901	0	0	0	0
20131101	2	0	4	6
20131201	0	0	1	1

Table 3. Monthly averages of measured solar radio bursts 2001-2013 for each class. Also total monthly averages are listed.

Year	Class I events	Class II events	Class III events	Monthly average
2001	0.4 (7.4%)	3.0 (55.6%)	2.0 (37.0%)	5.4
2002	2.0 (29.4%)	1.8 (26.5%)	3.0 (44.1%)	6.8
2003	0.9 (36.0%)	0.4 (16.0%)	1.2 (48.0%)	2.5
2004	0.8 (36.4%)	0.4 (18.2%)	1.0 (45.5%)	2.2
2005	2.0 (47.6%)	0.6 (14.3%)	1.6 (38.1%)	4.2
2006	0.3 (30.0%)	0.0 (0.0%)	0.7 (70.0%)	1.0
2007	0.0 (0.0%)	0.4 (57.1%)	0.3 (42.9%)	0.7
2008	0.0 (0.0%)	0.0 (0.0%)	0.0 (0.0%)	0.0
2009	0.0 (0.0%)	0.0 (0.0%)	0.0 (0.0%)	0.0
2010	0.0 (0.0%)	0.0 (0.0%)	0.3 (100.0%)	0.3
2011	0.8 (33.3%)	0.4 (16.7%)	1.2 (50.0%)	2.4
2012	0.7 (25.0%)	0.9 (33.3%)	1.1 (41.7%)	2.7
2013	0.4 (22.2%)	0.2 (11.1%)	1.3 (66.7%)	2.0

Table 4. Details about the observatories

Observatory	Frequency (GHz)	Location (degrees)	Size (m)
Metsähovi, Finland	11.2	60°N 24°E	1.8
Dominion, Canada	2.8	49°N 119°W	2.0
Learmonth, Australia	15.4	22°S 114°E	1.0
San Vito, Italy	8.8	41°N 18°E	2.4

From the beginning of year 2014 a new improved calibration method was introduced to allow a more accurate division of the solar radio bursts. The new calibration method enables the use of a realistic SFU values. In the future a new radome will be purchased, which will cancel out the most severe weather effects. The Northern location enables unique long-term monitoring during summertime.

Now new statistics are created, for example, with additional information of the shape for the burst. For example, in Figure 1 a double-peak structure is seen. Previously only the strength of the burst was taken into account even though double-peak structures were also measured. In the data measured in 2014 or prior a lot of bursts with a double-peak structure have been recorded. In the future these will be further studied. Solar radio bursts at 11.2 GHz will also be monitored and analyzed together with bursts measured with MRO Callisto equipment [11] , and also with RHESSI (ReuvenRamaty High Energy Solar Spectroscopic Imager) data [12] . In the near future 11.2 GHz data archive will be published for public use.

REFERENCES

1. Kallunki, J. (2009) Possibilities of the Mets?hoviRadiotelescopes for Solar Observations. Licentiate Thesis, Faculty of Information and Natural Sciences, Helsinki University of Technology, Espoo.
2. Khodachenko, M.L., Kislyakova, K.G., Zaqarashvili, T.V., et al. (2011) Possible Manifestation of Large-Scale Transverse Oscillations of Coronal Loops in Solar Microwave Emission. Astronomy and Astrophysics, 525, A105. http://dx.doi.org/10.1051/0004-6361/201014860
3. Kallunki, J., Lavonen, N., Jarvela, E. and Uunila, M. (2012) A Study of Long-Term Solar Activity at 37 GHz. Baltic Astronomy, 21, 255-262.
4. Benz, A.O. (2009) Quiet and Slowly Varying Radio Emissions of the Sun. The Landolt-Boernstein Database, LB VI/4B 4.1.1.6. http://dx.doi.org/10.1007/978-3-540-88055-4
5. Gary, D.E. and Keller, C.U. (Eds.) (2004) Solar and Space Weather Radiophysics. Astrophysics and Space Science Library, XXIV, 400 p.
6. Jiricka, K., Karlicky , M., Mészárosová, H. and Snízek, V. (2001) Global Statistics of 0.8-2.0 GHz Radio Bursts and Fine Structures Observed during 1992-2000 by the OndrejovRadiospectrograph. Astronomy and Astrophysics, 375, 243-250.http://dx.doi.org/10.1051/0004-6361:20010782
7. Nita, G.M. (2004) Statistical Study of Solar Radio Bursts. PhD Thesis, New Jersey Institute of Technology and Rutgers, The State University of New Jersey.
8. Natural Resources Canada. http://www.spaceweather.gc.ca/solarflux/sx-eng.php
9. US Air Force Weather Agency.http://www.afweather.af.mil/units/spaceweatheroperations.asp

10. NOAA's National Geophysical Data Center (NGDC). http://www.ngdc.noaa. gov/stp/space-weather/solar-data/solar-features/solar-radio/radio-bursts/reports/fixed-frequency-listings/
11. Kallunki, J., Uunila, M. and Monstein, C. (2013) Callisto Radio Spectrometer for Observing the Sun—Metsahovi Radio Observatory Joins the Worldwide Observing Network. IEEE Aerospace and Electronic Systems Magazine, 28, 5-9.http://dx.doi.org/10.1109/MAES.2013.6575404
12. RHESSI, ReuvenRamaty High Energy Solar Spectroscopic Imager.http://hesperia.gsfc.nasa.gov/rhessi2/

CITATION

Kallunki, J. and Uunila, M. (2014) Total Solar Flux Intensity at 11.2 GHz as an Indicator of Solar Activity and Cyclicity. *International Journal of Astronomy and Astrophysics*, **4**, 437-444. doi: 10.4236/ijaa.2014.43039.

Chapter 5

Developing an Advanced Prototype of the Acousto-Optical Radio-Wave Spectrometer for Studying Star Formation in the Milky Way

Alexandre S. Shcherbakov, Abraham Luna

National Institute for Astrophysics, Optics & Electronics (INAOE), Puebla, Mexico

ABSTRACT

The designed practically prototype of an advanced acousto-optical radio-wave spectrometer is presented in a view of its application to investigating the Milky Way star formation problems. The potential areas for observations of the cold interstellar medium, wherein such a spectrometer can be exploited successfully at different approximations, are: 1) comparison of the Milky Way case with extragalactic ones at scale of the complete galactic disk; 2) global studies of the Galactic spiral arms; and 3) characterization of specific regions like molecular clouds or star clusters. These aspects allow us to suggest that similar instrument will be really useful. The developed prototype of spectrometer is able to realize multi-channel wideband parallel spectrum analysis of very-highfrequency radio-wave signals with an improved resolution power exceeding 10^3. It includes the 1D-acousto-optic wide-aperture cell as the input device for real-time scale data processing. Here, the current state of developing this acousto-optical spectrometer in frames of the astrophysical instrumentation is briefly discussed, and the data obtained experimentally with a tellurium dioxide crystalline acousto-optical cell are presented. Then, we describe a new technique for more precise spectrum analysis within an algorithm of the collinear wave heterodyning. It implies a two-stage integrated processing, namely, the wave heterodyning of a signal in an acoustically square-law nonlinear medium and then the optical processing in the same solid-state cell. Technical advantage of this approach lies in providing a direct multi-channel parallel processing of ultra-high-frequency radio-wave signals with the resolution power exceeding 10^4. This algorithm can be realized on a basis of exploiting a large-aperture effective acousto-optical cell, which operates in the Bragg regime and performs the ultra-high-frequency co-directional collinear acoustic wave heterodyning. The general concept and basic conclusions here are confirmed by proof-ofprinciple experiments with the specially designed cell of a new type based on a lead molybdate crystal.

INTRODUCTION

Research in star formation is a keystone topic in Astrophysics. Decades of research in this field concluded that the stars are formed by a complex collapse process of the progenitor cloud, and modern models for the early formation inherent in clusters include now detailed descriptions of the accretion at disk and chemical implications [1] . The sites of star formation are normally "obscured" by the presence of material in those areas where stars are under formation, and this fact represents the major obstacle to improving our knowledge of the topic. Basic ideas and conclusions are inferred indirectly, and they require the data obtained from the corresponding models and/or new observations in various wavelength bands. One has to explore the spectral distributions of energy very accurately and within wide spectral ranges as is concluded from the elaborated numerical simulations [2] . Now we know that the maternal material is basically constituted by warm dust and cold molecular gas ($<100°K$). Due to a low extinction for long wavelengths, temperature, and composition of the medium around star forming regions, the most informative bands are infrared, microwave, and radio-wave ranges. The detailed phenomenology of star formation depends on the initial physical conditions of a region and combinations of these parameters accompanied by the cloud fragmentation, which includes magnetic, gravitational and velocity fields, and the accretion processes. These processes are not understood in detail, because we have not enough observational information on a widest set of environments. Fundamental parameters needed for a better understanding them include the mass, temperature, chemical composition, kinematics, magnetic field intensity, and density of gas and dust. These parameters are different for the regions of ongoing star formation, regions with potential of star formation, or old star formation regions. Observationally, these regions will form either low or high mass stars. Low mass stars (M < 8 Msun, in solar mass units) could form groups or be isolated, while high mass stars (M > 8 Msun) will be born only in groups. Questions related to: how, where, and when the star formation will evolve, represent basic branches of the researches. These researches could be theoretical and observational, galactic or extragalactic, and they could be related to the events occurring now or at the earliest stages of the Universe. The Initial Mass Function of stars (IMF), parameterizing the numbers of stars as a function of their mass, has an important influence on most of the observable properties of a galaxy, and detecting variations in the IMF could provide insights into the process of star formation. There are two main theories for the origin of the IMF: One is "competitive accretion" (stochastic) and the other one is "turbulent fragmentation" (deterministic, i.e the CMF is a precursor to the IMF). Since the IMF appears to be invariant, the deterministic theory concludes that the CMF must not depend much on environment. However, the CMF is poorly constrained at low masses (e.g. M <Msun) and there are some evidence for variations of the CMF in some regions, in such case the IMF may be finally determined or regulated by processes like accretion or feedback. Well studied regions of star formation in the 1kpc vicinity to the Sun are at the Gould Belt (c2d project and Hershel Gould Belt key project, [3] [4] , where a few star forming regions of high mass stars are placed. Then, in the Galactic context they have basically local characteristics and maybe a common origin [5] . The wide ranges for the above-mentioned physical parameters, which can be observed in the Milky Way now and along its evolution, produce the question about the contrast needed to detect the IMF variations expected for a CMF in the stochastic model. Generally, the galactic star formation could be analyzed using photometric or spectroscopic techniques. The observational group of the INAOE focuses

its attention mainly on young massive clusters and their interaction with its progenitor molecular cloud. Three following directions in researches of the star forming regions in the Milky Way context can be marked out first of all: 1) comparing the Milky Way case with extragalactic ones, for example, with M82 and M81; 2) global studies of the Milky Way spiral arms scales (about a few kpc); and 3) studies of specific regions like the Milky Way molecular clouds or star clusters.

The extragalactic approximation is basically comparisons of physical conditions towards the nearest prototypical starburst galaxy M82 and the normal spiral galaxy M81. We discover that M82 has spiral structure and hundreds of recently formed clusters of massive young stars [6] [7] . The normal spiral galaxy M81 also has hundreds of clusters, but contrasting with M82, not all of them are young and massive [8] . Comparison of M82 and M81 with the Milky Way imply that we are missing hundreds of young clusters in the Milky Way galaxy, and it opens an opportunity to research them. New instrumentation with better angular and spectral resolution and faster capabilities improves the correct classification and comparison between star formation scenarios on different kinds of galaxies [9]). In the case of M82 and M81, the usage of single dish antennas with large apertures has confirmed our previous results and improved our knowledge in the star forming process (Nobeyama 45 m and IRAM, [10] [11]). The ALMA interferometer at millimeter waves has spectrometers with thousands of km/s in bandwidth, and fractions of km/s in velocity resolution at sub-arcsecond of angular resolution, equivalently at 200 GHz: a bandwidth of 2 GHz with resolution ~1 MHz. With these capabilities the future comparison between Milky Way and extragalactic objects will be more effective: regions properly defined and variety of scenarios. The Alfonso Serrano Large Millimeter Telescope (LMT) also is starting its researches and will require instrumentation with similar capabilities to improve our knowledge of the kinematics and molecular chemistry in global galactic scenarios [12] .

The Galactic approximation to the star formation topic is using several now available galactic plane surveys in near and far infrared as well as in radio-wave and microwave bands. Recently, modern results have been reported to show the kinematics of the Milky Way disk as a characteristic behavior on the spiral arms. The behavior similar to a solid body rotation in the site of spiral arms, i.e. the normal site of star formation, implies lower angular momentum transferred to smaller scales and increases the efficiency of collapse of the molecular cloud [9] [12] [13] . Our privileged view of the detailed composition of the Milky Way also implies a difficulty in exploring its spiral structure and distinguishing correct environments and distances to objects and regions inside spiral structure [14] . It could be done through investigation of position and velocity in the phase space (i.e. on the velocity-position diagram). However, it requires spectral information basically obtained from radio and microwaves with several hundreds of km/s in bandwidth and fractions of km/s in velocity resolution at arc-minute of angular resolution, equivalently at 200 GHz: a bandwidth of 500 MHz and resolution of 60 kHz [15] [16] . The ALMA interferometer will be benefited of large-scale surveys that supply the basic information to go in deep spatial detail.

Specific galactic star forming regions are also studied in detail. Specially, we mention the regions where Young Massive Stellar Clusters (YMC), i.e. the objects missing in the Milky Way plane, should be born. There are no more of ten YMCs in the Milky Way plane detected, and they contain a large fraction of massive stars that evolve fast and died in supernova events. The calculated events regulate the life of a complete galactic disk through injecting the energy, transform the kinematical and chemical status of large regions (~100 pc), and could induce new star forming episodes [17] . The relation of YMCs with their interstellar medium around them is a pendent topic of research [3] . Westerlund 1 is the prototypical Galactic YMC and, it is under study to define its atomic and molecular environment and its possible production of gamma rays [18]). Star forming regions present physical characteristics and phenomena that are extremely actual processes of study: jets, molecular out flows, masers, extended green objects, and infrared dark clouds are nowadays part of the puzzle that give us clues to complete our understanding of the star forming process [19] -[21] . Stars in well-studied nearby low-mass star-forming regions (SFRs) follow the Kroupa initial mass function distribution (IMF). In a study of 12 highmass SFRs in their early embedded phase using the 24 μm source counts from Spitzer/MIPS, we find the distribution of the embedded sources in the associated molecular cloud follows the Kroupa IMF at the high-mass end, but shows a deficit of stars for masses lower than 2 Msun. The universality and shape of the IMF at the substellar mass range is still an investigation topic. In a study of 12 high-mass SFRs in their early embedded phase using the 24 μm source counts from Spitzer/MIPS, we have found that the distribution of the embedded sources in the associated molecular cloud follows the Kroupa IMF at the high-mass end, but shows a deficit of stars for masses lower than 2 Msun. On the other hand, there are extended 24 μm sources that are resolved into stars in the Spitzer/IRAC bands (~2"). Study of one such extended source surrounding the IRAS18236-1205 source, one of our 12 regions, would be able to spatially resolve this diffuse source into point sources, which would be analyzed to check whether they are the missing low-mass stars that are still in the process of formation. If so, these data would present serious challenge to the present ideas regarding the formation of low-mass stars [22] . All these processes require a high angular and spectral resolution from several hundreds of km\s in bandwidth and fractions of km\s in velocity resolution, equivalently at 200 GHz: a bandwidth of 500 MHz at resolution of 60 kHz, or also higher spectral resolution: a bandwidth of 50 MHz and resolution of 5 kHz at variable angular resolution from arcminutes to sub-arcseconds, now available with interferometers like ALMA [23] . The LMT with faster instrumentation and more adequate angular resolution should give us maps of the Milky Way plane at the microwave bands that will contribute to detecting the missing objects and its molecular environment with enough angular and spectral resolution.

Interdisciplinary collaboration in the astrophysics, optics, and electronics makes natural developing a new scientific instrumentation within the INAOE. Such projects like the LMT and HAWC (High Altitude Water Cherenkov observatory) require novel and sophisticated instrumentations to reach their goals. In connection with this, one can note the Aztec and Toltec, instruments for the LMT with bolometers at the wavelength 1 mm, the RSR (Redshift Search Receiver) with a spectrometer having total frequency bandwidth about 38 GHz and frequency resolutions of 30 MHz, which are equivalent to a bandwidth of thousands of km/s with a spectral resolution of decades of km/s, and the SEQUOIA spectrometer for images in the 100 GHz band with variable spectral resolution 0.1 - 10 km/s. In parallel, an advanced prototype of the acousto-optical spectrometer applicable to galactic maser emission at the frequency 43 GHz with the bandwidth about a few tens of km/s and a spectral resolution of fraction of km/s, which are equivalent to the frequency bandwidth of about 40 MHz with the frequency resolution close to 15 kHz is under construction in the INAOE [24] -[33] . Generally, acousto-optical spectrometers are used systematically, for example, in the satellite and airborne telescope instrumentation, like Herschel and SOFIA (Stratospheric Observatory for Infrared Astronomy), [34] , and in planet exploration like on Venus Express [35] . Their performances are at given quality and compactness together with a low energy.

GENERAL INTRODUCTIVE REMARKS

Here, an opportunity of developing an advanced prototype of the acousto-optic spectrometer for radio-astronomy is practically touched. The proclaimed advances mean that the presented approach intends an attempt of progressing in this area based mainly on involving new physical phenomena, front-rank algorithms for signal processing, and modern acousto-optical materials. In any prototype of similar spectrometer, a high-frequency radio-wave signal is injected into a large-aperture acousto-optical cell via a piezoelectric transducer. In fact, this signal produces numerous sets of dynamic acoustic gratings (each corresponds to a partial frequency contribution of the initial high-frequency radio-wave signal), which modulate the incident widely expanded light beam of a fixed wavelength. In so doing, the incident light beam is divided into a set of the corresponding partial beams arranging the independent parallel frequency channels and being scanned at the angles depending on the partial acoustic frequencies. The intensity of each partial light beam is determined by the amplitude of the corresponding partial frequency component from the initial radio-wave signal as well. Later, these partial light beams are focused by the Fourier transforming lens system on a linear CCD-array for the further computer processing. The optical layout of similar radio-wave spectrometer includes various optical components, namely, the powerful single-frequency laser, optical attenuator, light polarization controller, optical beam expander, acousto-optical cell (AOC), integrating lens, and CCD linear array, see Figure 1.

At this stage of development, the main goals assume the progress in characterization of these components and the estimation of potential performances peculiar to this prototype as a whole. During the characterization one can identify here optical sub-systems like, in particular, the precise beam shaper containing the laser, optical attenuator, polarization controller, and a multi-prism beam expander. These parts require really accurate adjusting to provide the incident light beam with the needed linear, circular or elliptical polarization in an expanded light beam. The processing sub-system of spectrometer consists of a triplet of the basic components. The first of them is represented by the AOC with an appropriate active optical aperture. The second component is the achromatic doublet lens (for example, from Thorlabs or Edmund Optics), while the third one is the CCD linear array (for instance, LC1-USB CCD, Thorlabs, numbering 3000 pixels of 7 μm × 200 μm each). The characterization of this sub-system is oriented to realizing sufficiently optimal resolution of a pattern determined by the sampling theorem requirements. A general view of an advanced prototype of the acousto-optic radio-wave spectrometer is presented on Figure 2.

The AOC is a key component of the spectrometer under consideration, because just this component determines the efficiency of operation, frequency bandwidth of spectrum analysis, and frequency resolution, i.e. the accuracy of radio-wave signals identification. Together with this, the AOC provides an opportunity to realize a multi-channel parallel processing on ultra-high carrier frequencies being perfectly adequate to the current needs of radio-astronomy explained in Section 1. These needs are widely varied from the frequency bandwidth $\Delta f = 40\,\text{MHz}$, the frequency resolution $\delta f = 15\,\text{kHz}$, the number of parallel signal processing (or, what is the same, the number of resolvable spots) $N \approx 2.7 \times 10^3$ and $\Delta f = 2.0\,\text{GHz}$, $\delta f = 1.0\,\text{MHz}$, $N = 2 \times 10^3$, which can be potentially realized at the top of modern technology exploiting, for example, specially designed AOCs based tellurium dioxide (TeO$_2$) and rutile (TiO$_2$) crystals, respectively, to more specific cases. These cases include wide variations for the needed performances as well, in particular, from $\Delta f = 500\,\text{MHz}$, $\delta f = 60\,\text{kHz}$, $N = 8.3 \times 10^3$ to $\Delta f = 50\,\text{MHz}$, $\delta f = 5.0\,\text{kHz}$, $N = 10^4$, which cannot be evidently realized within traditional AOCs and require principally novel approach to designing the AOCs considered below. This is why we touch here the possibility of improving the frequency resolution within parallel acousto-optical spectrum analysis via involving an additional nonlinear

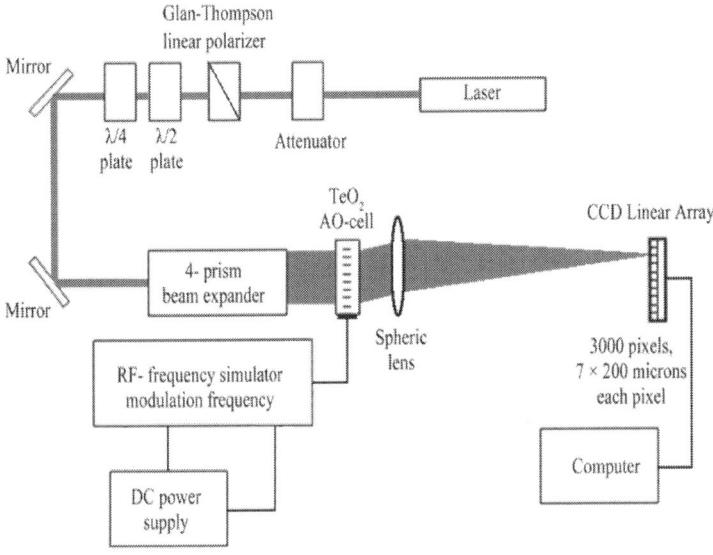

Figure 1. Schematic arrangement of an advanced prototype of the acousto-optical radio-wave spectrometer.

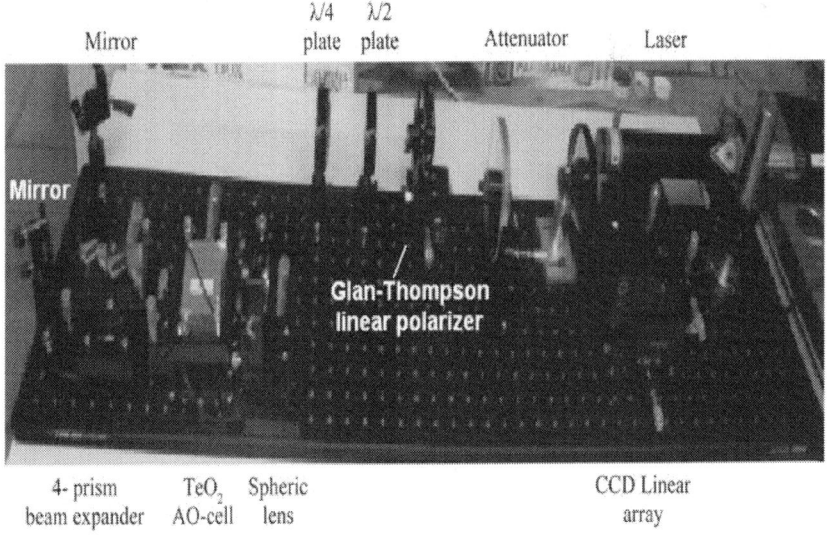

Figure 2. Photo of an advanced prototype of the acousto-optic radio-wave spectrometer.

namely, the wave heterodyning, into the data processing. The nonlinear process of wave heterodyning is realized through co-directional collinear interaction of the longitudinal acoustic waves of finite amplitudes. This process allows us either to improve the frequency resolution of spectrum analysis at a given frequency range or to increase by a few times the current frequencies of radio-wave signals under processing. Our theoretical and experimental findings are aimed at creating a new type of AOCs, which are able to improve the resolution inherent in acousto-optical spectrum analyzer operating over ultrahigh-frequency radio-wave signals. In particular, the possibility of upgrading the frequency resolution through the acoustic wave heterodyning is experimentally demonstrated using the AOC made of rather effective lead molybdate tetragonal single crystal. The obtained results show practical efficiency of the novel approach presented. Thereafter, potentials peculiar to the acousto-optical spectrum analysis of a gigahertz-frequency range radio-wave signals with essentially improved relative value of the frequency resolution, which can exceed the order of 10^{-4} in our case, are considered with exploiting a new type of the AOC.

A MULTI-PRISM BEAM EXPANDER

To realize a one-dimensional expanding of the laser beam one can exploit a set of rectangular prisms. Such a component is rather compact even with a large factor of expanding and can be done tunable in behavior. Using the well-known relation for light refraction [36] by the first border between air and glass $\sin\varphi = n\sin\delta$ (where n is the refractive index of a glass, α is the top angle, and φ is the angle of light incidence), one can obtain the factor of beam expanding

$$a)\ B_1 = \frac{d_1}{d_0} = \frac{\sqrt{\left(n^2 - \sin^2\varphi\right)\left[1 - n^2\sin^2\left(\alpha - \delta\right)\right]}}{n\cos\varphi\cos\left(\alpha - \delta\right)},$$

$$b)\ \delta = \arcsin\left(\frac{\sin\varphi}{n}\right).$$

$$(1)$$

In the simplest case, when all the prisms are identical to each other and the angles φ of incidence are the same for all of them, one can write $B_m = (B_i)^m$. If a number of prisms is even, the beam direction can be saved with an accuracy of some spatial parallel shift, see Figure 3. That is why the numbers $m = \{1,2,4\}$ are taken for consideration with the initial laser beam diameter of about $d_0 = 1\,mm$. In so doing, the needed optical aperture for an AOC can be chosen, while the corresponding angles φ of incidence become to be rather large. The plots illustrating light beam expansion by triplet of sets including $m = \{1,2,4\}$ prisms with $n = 1.5$ and $\alpha = 30°$, corresponding to the Littrow glass prisms, are presented by solid lines in Figure 4.

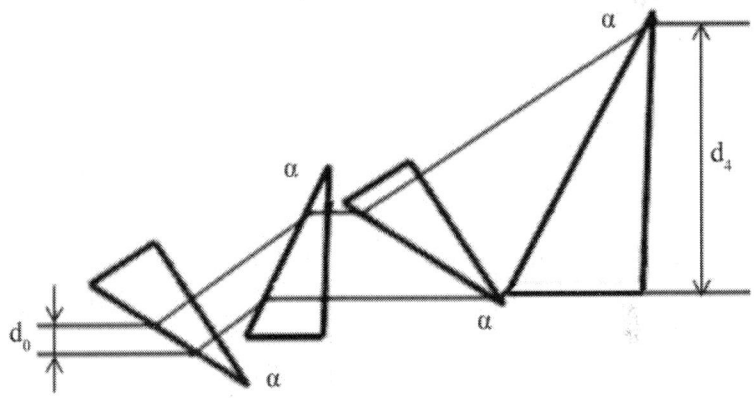

Figure 3. Passing the light beam through prisms: m = 4.

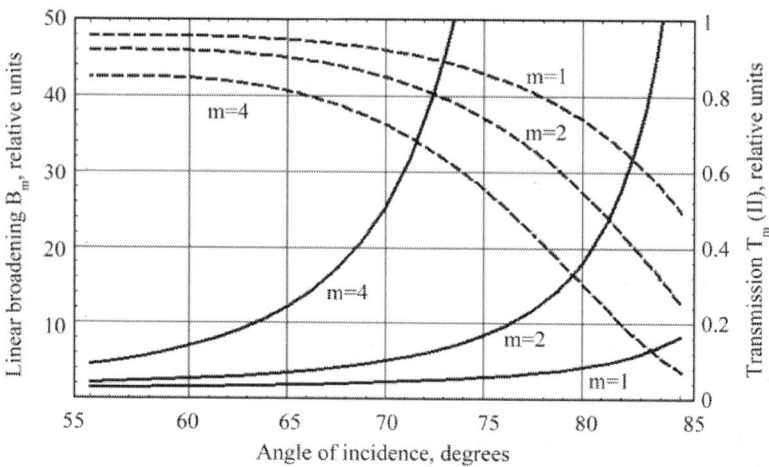

Figure 4. The combined diagram illustrating both the beam expanding and the transmission in glass prism shapers with m = {1, 2, 4}, n = 1.5, and α = 30°.

Then, one has to estimate the transmittance T of similar beam expander. For this purpose, one can adapt the well-known relations for transmittance during the refraction [36] , which depend on the state of light polarization. Two independent on each other states can be recognized, namely the state of polarization being orthogonal to the plane of incidence and, consequently, expanding a beam and the state of polarization belonging the plane of incidence. With these two options, one can write for each border between air and prism material that

$$a)\ T\left(\perp\right)=\frac{\sin 2\theta_i \sin 2\theta_t}{\sin^2\left(\theta_i+\theta_t\right)},$$

$$b)\ T\left(\|\right)=\frac{\sin 2\theta_i \sin 2\theta_t}{\sin^2\left(\theta_i+\theta_t\right)\cos^2\left(\theta_i-\theta_t\right)},$$

(2)

where θ_i and θ_t are the angles of incidence and transmittance at each individual prism facet. To describe the transmittance of one prism with, naturally, two facets, one should write

$$T_1(\perp) = \left[\frac{\sin 2\varphi \sin 2\delta}{\sin^2(\varphi+\delta)}\right]\left[\frac{\sin 2(\alpha-\delta)\sin 2\gamma}{\sin^2(\alpha-\delta+\gamma)}\right]$$

(3)

$$T_1(\|) = \left[\frac{\sin 2\varphi \sin 2\delta}{\sin^2(\varphi+\delta)\cos^2(\varphi-\delta)}\right] \times \left[\frac{\sin 2(\alpha-\delta)\sin 2\gamma}{\sin^2(\alpha-\delta+\gamma)\cos^2(\alpha-\delta-\gamma)}\right]$$

(4)

where $\gamma = \arcsin[n\sin(\alpha-\delta)]$. One can estimate that $T_1(\perp) < T_1(\|)$ always and,

consequently, $T_m(\perp) < T_m(\|)$ where $T_m(\perp) = [T_1(\perp)]^m$ and $T_m(\|) = [T_1(\|)]^m$.

Therefore, only the case related to $T_m(\|)$ will be considered later. Restricting ourselves by the particular case of laying the light polarization in the beamexpanding plane, we arrive at the following combined diagram for both the beam expanding and the optical energy transmission, which is shown in Figure 4 and looks rather convenient practically. The plots for the optical energy transmission in Figure 4 are governed by $T_m(\|)$ for $m = \{1, 2, 3, 4\}$. This combined diagram gives various practical possibilities. For instance, selecting the beam expansion factor of about $B = 35$, one can find the two options:

1) $m = 2$, $\varphi_2 \approx 83°$, $B_2 \approx 35$, and $T_2 \approx 0.4$ (i.e. 40%) or

2) $m = 4$, $\varphi_4 \approx 72°$, $B_4 \approx 35$, and $T_4 \approx 0.7$ (i.e. 70%), as they follow from the combined diagram in Figure 4.

The top angle α of glass prisms can be also optimized. Again, we take $m = \{2, 4\}$ and $n = 1.5$ for the same case of laying the light polarization in a beam expanding plane and find the contribution of the top angle α. Analysis shows that the influence of this angle is not too much, but it can be considerable in a view of precise optimization of the prism expander performances. Taking into account the performed analysis, a quartet $(m = 4)$ of the fused silica based BK-7 glass Littrow prisms with $n \approx 1.5$ and $\alpha = 30°$ can be chosen for a one-dimensional expansion of a laser beam from $d_0 = 1\ \text{mm}$ to $d_4 = 35\ \text{mm}$, providing each individual expansion at a value of $B_1 = 2.4323$, see Figure 2. The initial non-uniformity of lighting the input window of that AOC had been checked, and for this purpose the incident light intensity distribution had been fixed via photodetecting the beam profile obtained at the output facet of a 4-prism beam expander. The corresponding digitized oscilloscope trace for this distribution is depicted in Figure 5, so that one can estimate the non-uniformity (flatness) of its intensity by about 14%. The experimental estimation of transmittance has shown the value of 60%.

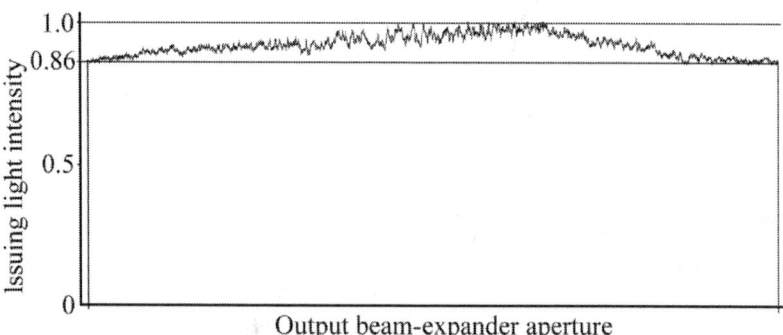

Figure 5. Experimental CCD-plot for the expanded beam profile, which represents the beam profile at the output facet of a 4-prism beam expander or what is the same, as the incident light distribution at the input facet of AOC.

CHARACTERIZATION OF THE FOURIER TRANSFORM OPTICAL SYSTEM

Resolvable Spot Characterization

Let us consider the Fraunhofer diffraction of the light with the wavelength λ (in a medium where the effect of Fraunhofer diffraction takes place) by a rectangular aperture, whose sizes are D_x and D_y. When the incident optical field amplitude $U_1(x_1, y_1)$ is uniformly distributed along this rectangular aperture, the pattern of Fraunhofer diffraction includes two sets of dropping maxima in two orthogonal to each other direction, so that each set of maxima can be analyzed individually [37] . This is why one can consider, for example, the light distribution in the x_f-direction lying in the plane $y_f = 0$. The zero-level width of the main lobe, i.e. the distance between the first two zeros, is $\Delta x_f = 2\lambda F/D_X$. The obtained estimation for Δx_f makes it possible to touch the optical resolution in acousto-optics. If resolution is restricted only by diffraction, the Rayleigh resolution criterion is usually exploited [36] . Within this criterion, two neighboring spots can be resolved if the intensity maximum of the first spot coincides with the first zero of the other one. Under such a condition, the minimal resolvable distance is given by $\delta x_f = \lambda F/D_X$. When just non-coherent light is in operation, the two identical spots are crossing each other at an intensity level of 0.405. The accuracy of aligning aperture of the CCD-linear array relative to the lens is practically important parameter. To find it one has to estimate the depth of the lens focus W, which can be approximately determined as $W = \pi(\delta x_f)^2/(2\lambda)$. Broadly speaking, this formula is exactly valid only for the focus depth belonging to the Gaussian beam [38] . Here, however, the focal depth W is applied to rectangular beam profile with some dropping at edges, which has been estimated as small enough. Then, taking into account the Rayleigh criterion and putting the size d_S of an individual resolvable spot can be estimated by δx_f in the diffraction limit, one can write $d_S = \lambda F/D$, where D_X is replaced by the AOC's aperture D. In an ideal optical system, the lighted length L_C in focal plane of the integrating lens can be expressed as $L_C = d_P N_P = d_S N_S$, where d_P and N_P are the size of an individual pixel and the number of the lighted pixels of a photo-detector, while N_S is the number of resolvable spots provided by the acousto-optical cell. When an individual resolvable spot lights more that one pixel, one can introduce the number m of pixels lighted by a one spot, which is $m = L_C/(d_P N_S) = d_S/d_P = N_P/N_S$. Finally, the lighted length L_C can be excluded, while the number m can be included, so that $F = D d_P m/\lambda$.

However, several things limit the performances of real optical system, and one of the most important factors is lens aberrations. Usually, optical systems operate within the paraxial approximation when $\sin\theta \approx \theta$ is reasonably valid for the angle of diffraction θ close to zero. With more highly curved surfaces, paraxial theory yields increasingly large deviations from real performances, because $\sin\theta \neq \theta$ at large angles of diffraction; and these deviations are known as aberrations. The aberrations are a measure of how the image differs from the paraxial prediction. Seidel [36] addressed this issue to contributions from $\sin\theta \approx \theta - (\theta^3/3!) + \cdots$, resulting in the third-order lens aberrations. To take into account the effect of aberrations in the first approximation one can formally substitute the aperture size D by some effective value $D_A = D/\vartheta_A$, which corresponds now to the perturbed image, where the factor $\vartheta_A > 1$ reflects the effect of aberrations. Together with this, the concrete magnitude of D_A can be found from the ray tracing procedure, which gives us an opportunity to determine first of all the individual spot size d_A conditioned by aberrations, so that one has to modify previous formulas with $\vartheta_A = d_A/d_S$. Under action of aberrations, the minimal resolvable distance δx_A , the size d_A , and the number m_A of pixels lighted by a spot, and the focus depth W_A take the forms: $\delta x_A = d_A = \lambda F/D_A$, $m_A = d_A/d_P$, and $W_A = \pi (\delta x_A)^2/(2\lambda)$, respectively.

The second important factor limiting the system performances is connected with optical quality of the AOC perturbed by the input electrical radio-wave signal. This factor includes all potential imperfectness of material and technology, and it can be estimated only experimentally, but similar estimations will naturally include already existing contribution from lens aberrations. Hence, one has to introduce the other factor $\vartheta_T = d_T / d_S > 1$ of total perturbations, which leads to $D_T = D / \vartheta_T$, $\delta x_T = d_T = \lambda F / D_T$, $m_T = d_T / d_P$, and $W_T = \pi (\delta x_T)^2 / (2\lambda)$.

Polarization Control

According to Ref. [36], it is principally possible to control the state of light beam polarization completely using only a pair of retardation plates, namely, half and quarter wave plates. Various lasers providing a single-mode continuous-wave radiation with linear polarization ratio about 100:1 and really low level of optical noise at a visible range have been used. For practical purposes, the intensity of light beam is regulated by a pair of linear optical attenuators, which can in principle affect the state of polarization. This is why an additional crystalline Glan-Thompson polarizer has been exploited thereupon to minimize possible variations from just linear polarization state of this source down to the ratio 100,000:1. A half-wave plate can rotate the initially linear polarization state by an arbitrary angle; for example, to the horizontal polarization state, which has been chosen as optimally matched with the selected laboratory coordinates. Then, a quarter-wave plate in combination with that half-wave plate allows us changing the polarization state from linear to arbitrary elliptical state and also controlling the angle of the ellipse of polarization. Potentially, rotating both the half and quarter wave plates makes it possible to control the polarization state even with the contribution of light-beam expander due to when the light beam passes through this expander, it changes the polarization state. All the measurements had been made using the polarimeter TX5004 (Thorlabs), which performs the needed measurements using the Poincare sphere technique.

EXPERIMENTAL DATA

Testing the optical system of an advanced prototype had been carried out with the Bragg cell, made of tellurium dioxide (TeO_2) crystal (Brimrose

Corp.), which has an active optical aperture of $35 \times 2\,\text{mm}$, see Figure 6(a). Within operating at the optical wavelengths of 532 or 633 nm with linear state of the incident light polarization on the central acoustic frequency of about 75 MHz, this cell provides the deflection angle of about 3 angular degrees and allows a maximum input acoustic power of about 1.0 W. The acoustic wave velocity can be estimated by $V \approx 0.65 \times 10^5\,\text{cm/s}$. The experimental studies consisted in two parts. The first one included measuring the bandwidths of acousto-optical interaction in the Bragg regime of light scattering in the first order. The second part of our experiments was related to estimating possible resolution of the AOC via measurements of the light intensity distributions of individual spots in focal plane of the integrating lens for light scattering by a TeO_2-cell in the first order. Figure 6(b) shows the experimental plot for the frequency bandshape inherent in the TeO_2-cell. One can observe the characteristic variations of efficiency at a top of the experimental plot. This oscillation is motivated by some uncoupling of both active and reactive parts of the cell's impedance at different frequencies. Each maximum of efficiency is potentially corresponding to better matching of impedance at the takes radiowave frequency. Practically obtained non-uniformity of the frequency characteristic is equal to about 14%. This non-uniformity is small enough to be easily compensated electronically within array camera when postprocessing the final bias response. Total experimental frequency bandwidth at a -3 dB-level has been estimated by $\Delta f_{exp} \approx 65\,\text{MHz}$.

To obtain sufficiently reliable estimations for the frequency resolution providing by the TeO_2–cell together with the above-described optical system (including the CCD-array with 3000 pixels of 7 μm ´ 200 μm each) a triplet of precise optical measurements had been performed at the wavelength 532 nm. The first of them has the goal of characterizing just the optical system without any contribution of an AOC. Therefore, at this step only an empty optical aperture of the needed sizes was lightedand the corresponding profile was registered. Figure 7(a) presents the obtained distribution in the focal plane of a lens. One can see that this profile corresponds to the spot size of about 16 μm with the side lobe level of about 5.15% for the first side lobe. The aim of the second measurement is: to take into account the quality of material used within

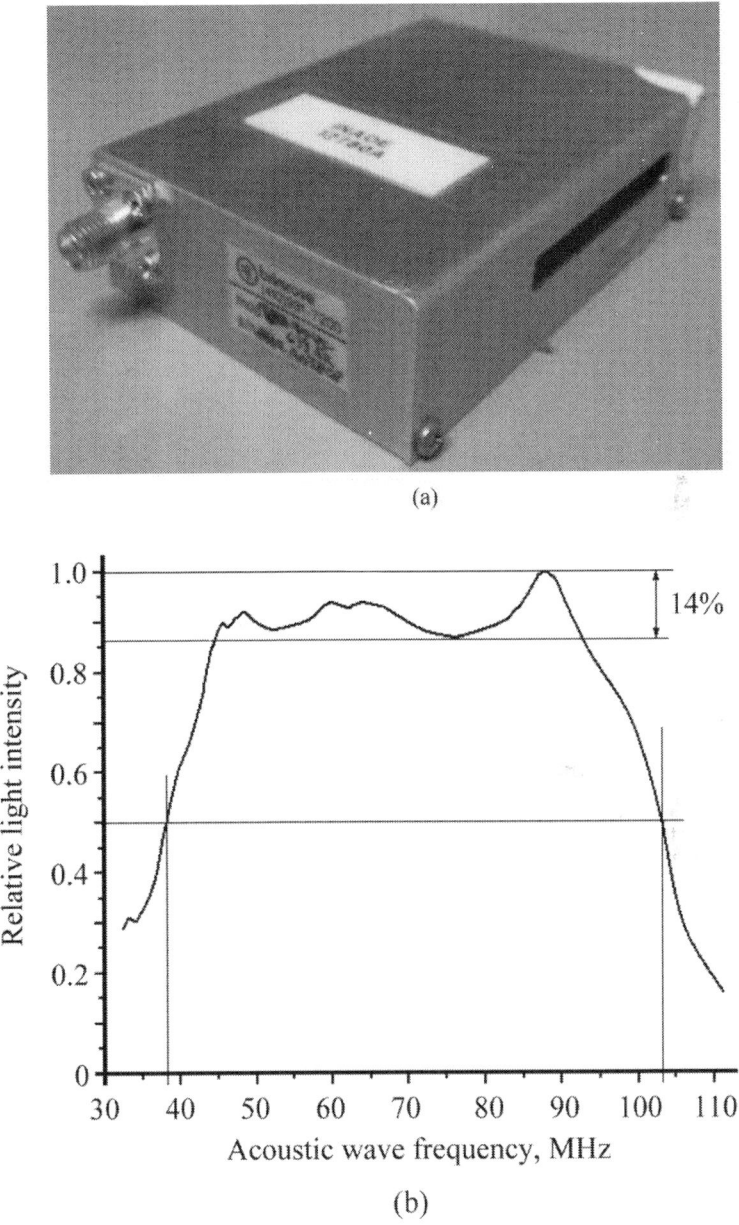

(a)

(b)

Figure 6. A 35-mm active aperture TeO₂-crystal based AOC, Brimrose Corp. (a) and the experimentally obtained frequency bandshape for this AOC (b, explanation in text).

manufacturing the AOC. This measurement should exploit real active optical aperture of the AOC, but without any electrical signal applied at the input port (the regime of so-called "cold cell"). Figure 7(b) demonstrates the light intensity profile gives the spot size of about 20 μm and the level of side lobes of about 6.0%. The third measurement had been realized in the regime of so-called "hot cell", i.e. with a radio-wave signal applied at the input port of TeO_2–cell. Figure 7(c) depicts the light intensity profile with the spot size of 21 μm and the side lobe level of about 6.3%.

Our experimental results had been obtained using the integrating lens # 30-976 (Edmund Optics) with $F = 85$ cm at the wavelength 532 nm, so that theoretically $d_S = \lambda F / D \approx 12.92\,\mu m$. Together with this, plots in Figure 7 exhibit $d_A \approx 16\,\mu m$ and $d_T \approx 21\,\mu m$, hence $\vartheta_A \approx 1.238$, $\vartheta_T \approx 1.625$, $m_T \approx 3$, and $D_T \approx 21.53$ mm. The last data show that about 38.5% of active optical aperture of the AOC is lost due to imperfectness of the lens and cell's material. Then, instead of theoretical limit of frequency resolution $\delta f = V / D \approx 18.6$ kHz, one yields the measured value $\delta f_T = V / D_T \approx 30.2$ kHz.

Thus the obtained number of resolvable spots is $N_{exp} \approx \Delta f_{exp} / \delta f_T \approx 2152$. Finally, the expected lighted length in focal plane of the integrating lens can be considered as $L_{CE} = d_T \cdot N_{exp} \approx 45.2$ mm. This length potentially includes about 6450 pixels of 7 μm each, i.e. capabilities of the optical system under consideration exceeded possibilities of the exploited CCD-array with 3000 pixels by more than two times.

FREQUENCY AND RESOLUTION PERFORMANCES OF THE ORDINARY BRAGG AOC

Let us estimate generally the frequency bandwidth Δf, the frequency resolution δf, and the number N of resolvable spots inherent in the

Bragg AOC operating in a one-phonon normal light scattering regime. The frequency bandwidth is given by [39] [40] $\Delta f \approx 2nV^2 / (\lambda f L)$. The Bragg limit is determined by the inequality for Klein-Cook parameter $Q = \lambda f^2 L / V^2 \geq 2\pi$ [41]. In this limit, the Heisenberg uncertainty principle leads to the frequency resolution $\delta f \approx V/D = T^{-1}$, where T is the time aperture of AOC with the space aperture D. The number N of resolvable spots is given by the ratio $N = \Delta f / \delta f = T \Delta f$. In ultra-high-frequency AOCs, the value of N is restricted by both the geometrical factors and the acoustic attenuation in cells' material. The first

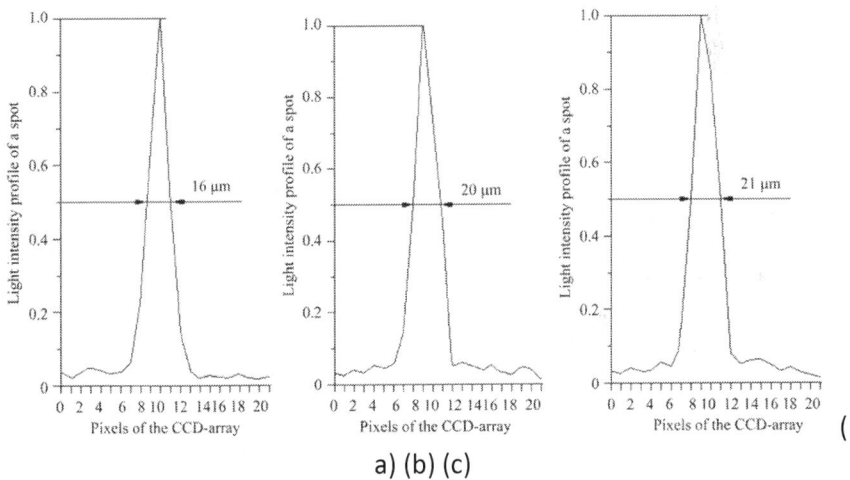

a) (b) (c)

Figure 7. Light intensity profiles of individual spots: (a) empty aperture without AOC, (b) AOC without the input radio-wave signal and c) AOC with the input radio-wave signal.

geometric factor is the maximal size of aperture D, so that one can estimate the maximal bandwidth as $\Delta f \approx f_0/2$, where f_0 is the central carrier frequency of the acoustic wave, and obtain the first limitation $N_1 \leq D f_0 / (2V)$. The second geometric factor is governed by the acoustic beam divergence. Let the aperture D belongs to the near zone of acoustic field radiated by the piezoelectric transducer, whose size is L, so that $D \approx L^2 f_0/(2V)$. This relation leads to $L = nV^2 Q/(2\pi\lambda f_0^2)$ and to the second limitation $N_2 \leq (nVQ)^2/(4\pi\lambda f_0)^2$. The third limitation is conditioned by acoustic attenuation being a function of f_0.

The level χ dB of attenuation allows the aperture $D \leq \chi\Gamma_0^{-1} f_0^{-2}$, where the factor Γ_0 of acoustic losses expressed in [dB/(cm×GHz^2)]. Substituting this formula into formula for N_1, one can find $N_3 \leq \chi/(2\Gamma_0 V f_0)$. Thus, the spot number N is restricted by a triplet of these independent limitations.

To make numerical estimations let us take a lead molybdate (PbMoO$_4$) crystalline AOC, exhibiting: $V = 3.63 \times 10^5$ cm/s, $\lambda = 633$ nm, n = 2.26, and $\Gamma_0 = 15$ dB/(cm·GHz^2) [39] [42] [43]. The numerical estimations have been realized for the apertures D = 1 - 4 cm; the attenuation factors along the total aperture $\chi = 4$ and 6 dB/aperture, and the Klein-Cook parameter $Q = 2\pi, 3\pi, 4\pi$, see Figure 8. It is seen that a lead molybdate AOC with $D \approx 2$ cm, $Q = 2\pi$, and $\chi < 4$ dB/aperture is capable to provide $N \approx 700$ with $\delta f \approx 180$ kHz within $\Delta f \approx 120$ MHz at a central frequency $f_0 \approx 250$ MHz. Together with this, using Figure 8 one can conclude that conventional lead molybdate deflector even with $D = 1$ cm is not operable at $f_0 \geq 600$ MHz.

Thus, now one can formulate the question facing this section. Taking alone a given lead molybdate AOC with the given aperture $D = 2$ cm, is it possible to keep the same number of resolvable spots with the same potential frequency resolution in the same frequency bandwidth at significantly increased central carrier frequency f_0 exceeding 600 MHz? The further consideration gives definitely positive answer to this question under condition of exploiting the collinear acoustic wave heterodyning technique.

OPERATING AN ACOUSTO-OPTICAL CELL WITH THE WAVE HETERODYNING

Let us overview potential possibilities related to exploiting a collinear wave mixing in a medium without any group-velocity dispersion while with strongly dispersive losses. This medium allows realizing effective wave heterodyning, when the beneficial data in the signal wave are converted from a relatively high-frequency carrier wave to a difference-frequency wave. The accuracy of frequency measurements is physically determined by the uncertainty in the energy or momentum inherent in a photon localized in the interaction area [44] . Due to rather strong dispersion of losses, the heterodyning leads to increasing the characteristic length and time of propagation (they both are associated with a clear optical aperture) for the converted signal in that medium and to improving significantly the accuracy of signal processing. As a result, the real-time optical analysis of frequency spectra, belonging to analogue ultra-high frequency radio-wave signals, can be done with considerably improved frequency resolution. This result is based on a two-cascade processing, i.e. on exploiting a pair of different wave processes one after the other sequentially in the same crystalline AOC. This cell includes two resonant piezoelectric transducers, converting the input electronic signals into gigahertz-frequency elastic waves, with the corresponding electronic ports on its upper facet, clear optical aperture D, and an effective acoustic absorber on its bottom facet, see Figure 9.

The first wave process represents mixing the longitudinal elastic waves of finite amplitudes in a compactly localized upper domain of a cell where relatively powerful pump of the frequency f_P interacts with relatively weaker signal elastic wave of the frequency f_S. Within this nonlinear process the collinear wave heterodyning takes place providing appearance of an elastic wave of the difference-frequency f_D, which is able to propagate along a large-aperture cell due to weaker manifestation of strongly dispersive losses at lower frequencies. The second wave process is the subsequent Bragg light scattering by the difference-frequency elastic wave in as possible linear regime, i.e. in the regime of a

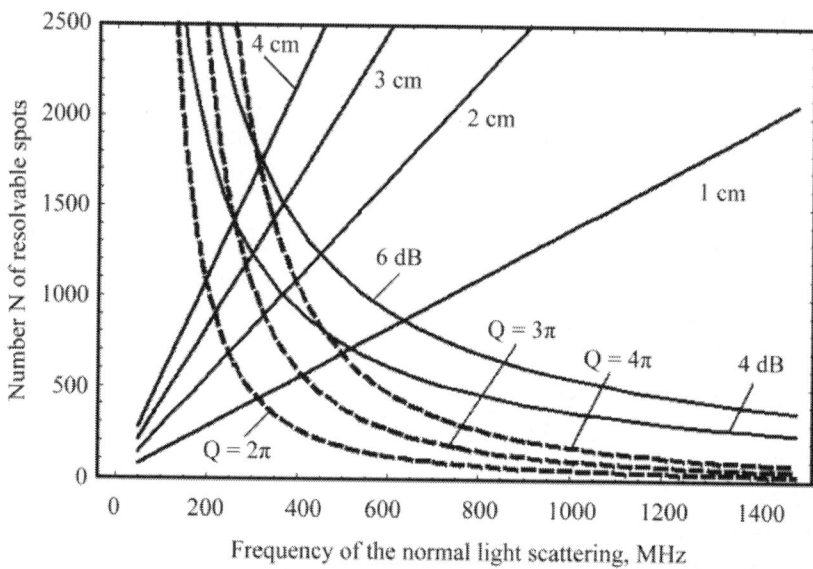

Figure 8. The combined effect of a triplet of the restricting factors. The solid straight lines are related to N_1 with the apertures D = 1, 2, 3, and 4 cm. The dashed lines regards to N_2 with Q = 2p, 3p, and 4p. The solid falling curves illustrate N_3, i.e. the acoustic attenuation with total losses of 4 and 6 dB along the optical aperture.

given acoustic field for the incident light beam. This process occurs within a clear aperture D lighted by a wide incident optical beam of the wavelength λ and is able to realize optical spectrum analysis by itself. When, for example, the signal wave is rather intricate in behavior and consists of various frequencies, each individual spectral component from the difference-frequency elastic wave plays the role of a partial thick dynamic diffractive grating for the incident light beam. The length L of interaction has to provide the Bragg regime of light scattering. In so doing, the acousto-optical spectrum analysis of a gigahertz-frequency range radio-wave signals with essentially improved frequency resolution can be realized. During our proof-of-principle experiments a new type of the acousto-optical cell made of rather effective lead molybdate single crystal was exploited. These experimental data show that the elaborated approach, based algorithmically on a two-cascade processing, allows a direct multi-channel parallel optical analysis of spectra inherent in ultra-high-frequency radio-wave signals at relative accuracy better than 10^{-4}.

PERFORMANCES OF A NOVEL LEAD MOLYBDATE CRYSTALLINE AOC

A novel solid-state acousto-optical cell had been designed and used in the standard scheme of a prototype for the acousto-optical spectrum analyzer of ultra-high-frequency radio-signals with the laser ($\lambda = 440\,nm$, the issuing optical power exceeds 100 mW), large-aperture achromatic doublet lens, and a 3000-pixel CCD linear array. A lead molybdate(PbMoO$_4$) single crystal$(\Gamma_0 = 15\,dB/cm \cdot GHz^2)$ with practically available optical aperture $D = 3.7\,cm$, oriented along the [001]-axis for an acoustic beam along [100]-axis for an optical beam [42] [43] , was used in that cell. The cell completed with a pair of electronic input ports for the pump and signal on one of its facets as well as with acoustic absorber on the opposite facet. This crystalline material was chosen because of its high value of the relative figure of acousto-optical merit that can be characterized by a value of $M_2 \approx 36.3 \times 10^{-18}\,s^3/g$ for both possible eigen-states of light polarization in this tetragonal crystal and its rather high acoustic interaction efficiency for collinear longitudinal waves in the [001]-direction described by $|\Gamma| = 17.5$ [42] [43] . The piezoelectric transducer with length of $L = 1.0\,cm$, generating the signal wave with power density of ~ 100 mW/mm^2, was made of a thin $(Y + 36°)$-cut lithium niobate, so that it excited purely longitudinal acoustic wave, with conversion losses of about 2 dB at its resonant frequency close to 1530 MHz. The singlefrequency pumping longitudinal acoustic wave with the power density of up to 600 mW/mm^2 was generated at a carrier frequency of approximately $f_p = 1390\,MHz$, so that the case of $\gamma \in [1.04, 1.20]$ had been experimentally realized. During the experiments, we have placed a diaphragm in about 5-mm vicinity of the

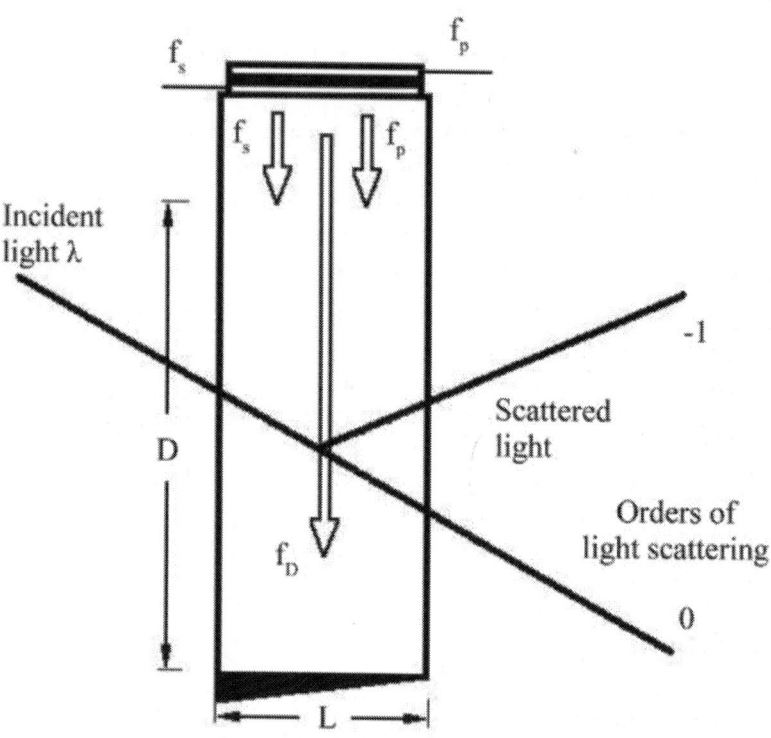

Figure 9. Schematic arrangement of the interacting beams in a two-cascade AOC [45] .

piezoelectric transducers area (about 13.5% of the total aperture) to minimize the effect of this area, where an increase in the power of difference frequency waves takes place. Consequently, the working optical aperture of a cell was a little bit longer than 3.7 cm. The bandwidth of that prototype was about 180 MHz. The efficiency of light scattering by an additional acoustic wave at the difference-frequency was slightly exceeding 1%. Figure 10 shows the digitized oscilloscope traces of amplitude-frequency distribution peculiar to that prototype with the acoustooptical cell based on the collinear wave heterodyning. The digitized trace of this distribution had been recorded by a multi-pixel CCD linear array photo-camera through connecting the input signal port of a cell at an ultrahigh-frequency radio-wave sweep-generator and fulfilling the acoustic wave heterodyning in a lead molybdate crystal. For a radio-wave signal, producing the dynamic acoustic grating on the resulting carrier difference-frequency of about 235 MHz, the attenuation is close to 3 dB over the total cell aperture. At the same time, for the signal acoustic waves at even the lower original frequency $f_s = 1440\,\mathrm{MHz}$ the attenuation exceeds 100 dB along a 3.7 cm aperture, which is perfectly unacceptable in practice. Within the second set of our experimental tests, we examined the resolution of spectrum analyzer with the cell exploiting the collinear acousticwave heterodyning. In fact, the intensity distribution of an individual resolvable spot in the focal plane of the integrating lens had been considered. In so doing, the technique, which had long been in use, with very narrow slit diaphragm scanning over sufficiently sensitive photodetector and subsequent logarithmic amplifier was applied to our needs. Practically, this technique gives an opportunity to fix the continuous distribution of light intensity in the lobes of an individual spot really carefully in a rather wide dynamic range of about 25 dB [45] [46] . One can see that the measured level of the first lobes lies at a level of about −13 dB with initially homogeneous lighting of the operating cell's aperture, which is in good coincidence with the well-known theoretical prediction [45] [46] . Figure 11 presents the digitized trace, which had been recorded in the focal plane of the integrating lens via applying exactly a single-frequency excitation at the input signal port of the proposed cell. In the case under consideration, physical limit of the frequency resolution is $V/D \approx 98\,\mathrm{kHz}$, while experimentally obtained value, affected by acoustic losses as well as by technical imperfectness of the integrating lens, corresponds to a frequency resolution of about $\delta f \approx 120\,\mathrm{kHz}$ at a level of −10 dB and gives the number of resolvable spots or, what is the same, the number of parallel frequency channels $N \approx 1500$. By this is meant that the proposed technique for direct parallel optical spectrum analysis of gigahertz-frequency range radio-wave signals provides at least 1500-channel processing even within our proof-of-principle experiment. As this takes place, the accuracy or the relative frequency resolution $\delta f / f_s$ (where $f_s = 1444\,\mathrm{MHz}$) is better than 10^{-4}, which is practically unattainable for conventional direct acousto-optical methods of spectrum analysis.

BRIEF DISCUSSION AND CONCLUSIVE REMARKS

The presented information makes it possible to have a look at the potential progress within designing modern prototype of the acousto-optical spectrometer for precise analysis of various radio-wave signals in radio-astronomy. The chosen arrangement of a prototype realizes the direct analysis of radio-wave signal in frequency plane of the Fourier transform system. Moreover, such an approach allows in the future significant schematic development based on exploiting new physical principles as well as involving novel acousto-optical materials being capable to improve the performance data of spectrometer. The obtained practical results confirm the validity of the chosen technical solutions as a whole. Nevertheless, it is clearly seen that, for instance, quality of some mechanical and optical components, in particular the lens exploited with the peculiar mechanical mount, should be substantially increased to provide a great help in improving performances of the prototype developed. Then, it should be noted that experimental technique of measurements and characterization has to be also progressed.

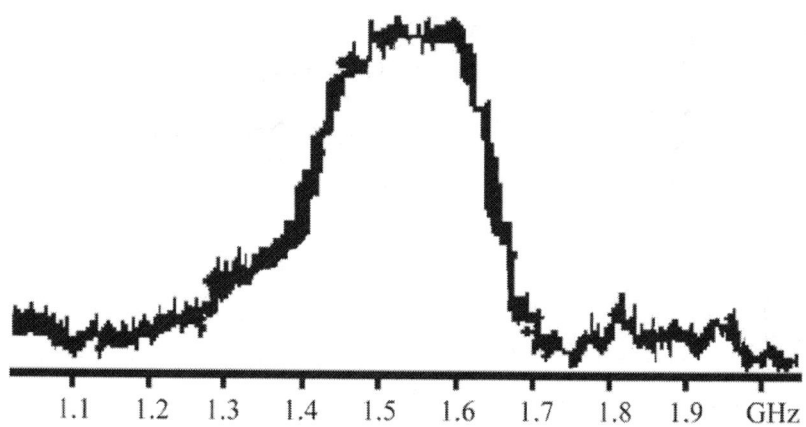

Figure 10. The digitized oscilloscope trace of the bandshape of the PbMoO4 crystaline AOC with collinear heterodyning exploiting the longitudinal elastic waves of finite amplitude in a crystal [46] .

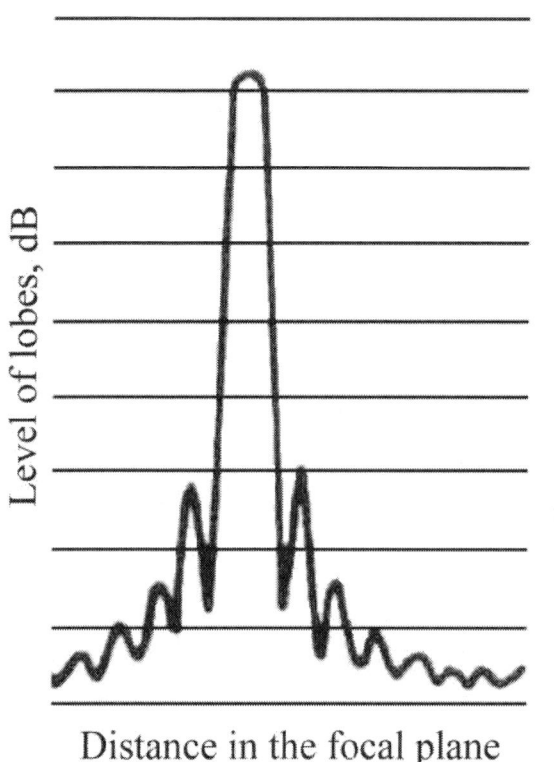

Distance in the focal plane

Figure 11. The digitized profile of an individual resolvable spot peculiar to a PbMoO4-cell with the most uniform distribution $(f_s/f_p \approx 1.14)$ of difference-frequency wave distributed along 3.7 cm aperture; horizontal lines are spaced by 2.5 dB [46] .

Together with this, we demonstrate both the possibility and the potential advantages of applying a co-directional collinear wave heterodyning to essential, about an order of magnitude, improvement of the frequency resolution within a multi-channel parallel acousto-optical spectrum analysis of gigahertz-frequency range analogue radio-wave signals. In so doing, an opportunity of implementing acousto-optical data processing with the wave heterodyning has been experimentally performed utilizing the specially designed acousto-optical cell made of a lead molybdate single crystal. The proposed technique exploits a two-cascade algorithm of processing and is intended for direct parallel and precise optical spectrum analysis and provides about 1500-frequency-channels for processing analogue radio-wave signals in a gigahertz-frequency range with the accuracy or, what is the same, with the relative frequency resolution better than 10^{-4}, which is usually unattainable for conventional direct acousto-optical methods. The obtained results reflect fruitful character of modern approaches based on applying various non-linear phenomena to improve the performance data of optical processing and give an appropriate example of this kind. At the moment, a few practical advantages of the presented approach can be noted. First, the proposed device does not need additional electronic equipment for mixing the signals and selecting the resulting currier frequency, because heterodyning can be performed directly in a cell and provide potentially the dynamic range of about 90 dB peculiar to wave processes in solids. Then, the approach under consideration decreases the required relative bandwidth of piezoelectric transducer from $50\% - 100\%$ at the resulting frequency within a conventional cell to $10\% - 15\%$ at the initial carrier frequency. Within our proof-of-principle experiment the acousto-optical cell with 2 piezoelectric transducers was used, but generally it is not necessary. Due to the relative bandwidth does not exceed 15%, potentially it is quite reasonable to exploit just only one transducer. Third, in the case of a spatially multi-channel arrangement of the acousto-optical cell, the identity of neighboring spatial channels to each other can be provided by adjusting the corresponding heterodynes. Finally, one should note that the number of isotropic or crystalline materials, which are appropriate for acousto-optical cells processing signals in a gigahertz-frequency range, is definitely restricted due to fast-growing influence of a square-law frequency dependence for the acoustic attenuation in solids. For instance, one can easily show [45] [46] that the above-discussed lead molybdate crystal cannot be used for creating a conventional acousto-optical cell operating with signals whose carrier frequency exceeds about 300 - 400 MHz. Nevertheless, just this crystalline material had been exploited for the control over 1.5 GHz signals within these studies. Consequently, one can conclude that a two-cascade arrangement of a cell presented here allows extending the spectrum of acousto-optical materials being appropriate for direct processing of ultra-high-frequency analogue radiowave signals. Remarkable qualities of the acousto-optical spectrometers are stability, compactness and low energy consumption, such qualities now could be potentiated for using the collinear wave heterodyning discussed in this document, because this new characteristic gives us the opportunity to explore the dynamic selection of frequency bandwidth and resolution using the acousto-optical spectroscopy technique.

ACKNOWLEDGEMENTS

This work has been financially supported by the CONACyT, Mexico within the projects # 61237 (initially) and 182841.

REFERENCES

1. Bastian, N., Cabrera-Ziri, I., Davies, B. and Larsen, S.S. (2013) Constraining Globular Cluster Formation Through Studies of Young Massive Clusters—I. A Lack of Ongoing Star Formation within Young Clusters. Monthly Notices of the Royal Astronomical Society, Advance Access, 12. http://dx.doi.10.1093/mnras/stt1779

2. Offner, S., Robitaille, T.P., Hansen, C., McKee, C.F. and Klein, R.I. (2012) Oberving Simulated Protostars with Outflows: How Accurate Are Protostellar Properties Inferred from SEDs?" Astrophysical Journal, 753, 98-115. http://dx.doi.10.1088/0004-637X/753/2/98

3. Il Evans, N.J., Dunham, M.M., Jørgensen, J.K., Enoch, M.L., Merín, B., van Dishoeck, E.F., Alcalá, J.M., Myers, P.C., Stapelfeldt, K.R., Huard, T.L., Allen, L.E., Harvey, P.M., van Kempen, T., Blake, G.A., Koerner, D.W., Mundy, L.G., Padgett, D.L. and Sargent, A.I. (2009) The Spitzer c2d Legacy Results: Star-Formation Rates and Efficiencies; Evolution and Lifetimes. Astrophysical Journal Supplements, 181, 321-350.http://dx.doi.10.1088/0067-0049/181/2/321

4. Ph. André, et al. (2010) From Filamentary Clouds to Prestellar Cores to the Stellar IMF: Initial Highlights from the Herschel Gould Belt Survey. Astronomy & Astrophysics, 518, L102-L109. http://dx.doi.10.1051/0004-6361/201014666

5. Grenier, I.A. (2004) The Gould Belt, Star Formation, and the Local Interstellar Medium. arXiv:astro-ph/0409096.

6. Mayya, Y.D., Carrasco, L. and Luna, A. (2005) The Discovery of Spiral Arms in the Starburst Galaxy M82. Astrophysical Journal, 628, L33-L36.http://dx.doi.10.1086/432644

7. Mayya, Y.D., Romano, R., Rodríguez Merino, L., Luna, A., Carrasco, L. and Rosa González, D. (2008) HST ACS Imaging of M82: A Comparison of Mass

and Size Distribution Functions of the Younger Nuclear and Older Disk Clusters. Astrophysical Journal, 679, 404-419. http://dx.doi.10.1086/587541

8. Mayya, Y.D., Rosa González, D., Santiago-Cortés., M., Arellano-Córdova, K. and Rodríguez, M. (2013) GTC Long-Slit Spectroscopy of Compact Stellar Clusters in M81. Revista Mexicana de Astronomía y Astrofísica Conference Series, 42, 22-23.

9. Kennicutt, R.C. and Evans, N.J. (2012) Star Formation in the Milky Way and Nearby Galaxies. Annual Review of Astronomy & Astrophysics, 50, 531-608.http://dx.doi.10.1146/annurev-astro-081811-125610

10. Salak, D., Nakai, N., Miyamoto, Y., Yamauchi, A. and Tsuru, T.G. (2013) Large-Field CO(J = 1 → 0) Observations of the Starburst Galaxy M 82. Publications of the Astronomical Society of Japan, 65, 66-81.

11. Heithausen, A. (2012) On the Nature of Dust Clouds in the Region towards M81 and NGC 3077. Astronomy & Astrophysics, 543, 21-27. http://dx.doi.10.1051/0004-6361/201117861

12. Elmegreen, B.G. (2011) Star Formation on Galactic Scales: Empirical Laws. European Astronomical Society Publications Series, 51, 3-17. http://dx.doi.10.1051/eas/1151001

13. Luna, A., Bronfman, L., Carrasco, L. and May, J. (2006) Molecular Gas, Kinematics, and OB Star Formation in the Spiral Arms of the Southern Milky Way. Astrophysical Journal, 641, 938-948. http://dx.doi.10.1086/500163

14. Efremov, Yu.N. (2011) On the Spiral Structure of the Milky Way Galaxy. Astronomy Reports, 55, 108-122. http://dx.doi.10.1134/S1063772911020016

15. Bronfman, L., May, J. and Luna, A. (2000) A CO Survey of the Southern Galaxy. In: Mangum, J.G. and Radford, S.J.E., Eds., Imaging at Radio through Submillimeter Wavelengths, Astronomical Society of the Pacific Conference Series, 217, 66-71.

16. Sawada, T., Hasegawa, T., Handa, T., Morino, J.-I., Oka, T., Booth, R., Bronfman, L., Hayashi, M., Luna Castellanos, A., Nyman, L.-A., Sakamoto, S., Seta, M., Shaver, P., Sorai, K. and Usuda, K.S. (2001) The Tokyo-Onsala-ESO-Calan Galactic CO J = 2-1 Survey. I. The Galactic Center Region. Astrophysical Journal Supplements, 136, 189-219.http://dx.doi.10.1086/321793

17. Mayya, Y.D., Luna, A., Carrasco, L. and Bronfman, L. (2012) The Interplay Between the Young Stellar Super Cluster Westerlund 1, and the Surrounding Interstellar Medium. European Physical Journal Web of Conferences, 19, 8006- 8009.http://dx.doi.10.1051/epjconf/20121908006

18. Luna, A., Mayya, Y.D., Carrasco, L. and Bronfman, L. (2010) The Discovery of a Molecular Cavity in the Norma Near Arm Associated with H.E.S.S gamma-ray Source Located in the Direction of Westerlund 1. Astrophysical Journal, 713, L45-L49.http://dx.doi.10.1088/2041-8205/713/1/L45

19. Zinnecker, H. and Yorke, H.W. (2007) Toward Understanding Massive Star Formation. Annual Review of Astronomy and Astrophysics, 45, 481-563.http://dx.doi.10.1146/annurev.astro.44.051905.092549
20. Retes, R., Luna, A., Mayya, D. and Carrasco, L. (2009) Embedded Young Stellar Population in the Molecular Region Towards IRAS 18236-1205 Source. Revista Mexicana de Astronomía y Astrofísica, Conference Series, 37, 165-169.http://adsabs.harvard.edu/abs/2009RMxAC..37..165R
21. Retes, R., Luna, A., Mayya Y.D. and Carrasco, L. (2011) Characterizing the Embedded Young Stellar Objects in the Galactic Star-forming Region IRAS 18236-1205. Revista Mexicana de Astronomía y Astrofísica, Conference Series, 40, 249-250.http://adsabs.harvard.edu/abs/2011RMxAC..40..249R
22. André, Ph., Könyves, V., Arzoumanian, D., Palmeirim, P. and Peretto, N. (2013) Star Formation as Revealed by Herschel. Astronomical Society of the Pacific Conferences, 476, 95.
23. Bronfman, L. and Merello, M. (2013) From Large Scale Surveys of the Galaxy to High Resolution Observations with ALMA. Astronomical Society of the Pacific Conferences, 476, 231.
24. Shcherbakov, A.S., Balderas Mata, S.E., Tepichin Rodriguez, E., Luna Castellanos, A., Sanchez Lucero, D. and Maximov, Je. (2007) The Main Peculiarities of Arranging the Optical Scheme of Acousto-Optical Spectrometer for the Mexican Large Millimeter Telescope. Proceedings of SPIE, 6663, 1-9.
25. Shcherbakov, A.S., Luna Castellanos, A., Balderas Mata, S. E. and Maximov, Je. (2007) Upgrading the Frequency Resolution of Spectrum Analyzers for Radio-Astronomy due to Exploiting a Multi-Phonon Light Scattering in TeO_2 Crystalline Modulators. Proceedings of SPIE, 6796, 1-10.
26. Shcherbakov, A.S., Luna Castellanos, A. and Balderas Mata, S.E. (2007) Optical Modulators Exploting a Multi-Phonon Light Scattering in TeO_2 Structures. Proceedings of SPIE, 6796, 1-12.
27. Shcherbakov, A.S., Tepichin Rodriguez, E., Aguirre Lopez, A. and Maximov, Je. (2009) Frequency Bandwidth and Potential Resolution of Optical Modulators Exploiting a Multi-Phonon Light Scattering in Crystals. Optik—International Journal for Light and Electron Optics, 120, 301-312. http://doi:10.1016/j.ijleo.2007.06.027
28. Herrera Martinez, G., Luna Castellanos, A., Carrasco Bazúa, L., Shcherbakov, A.S., Sanchez Lucero, D. and Mendoza Torres, E. (2009) A Design of an Acousto-Optical Spectrometer. Mexican Review of Astromomy& Astrophysics, Conference Series, 37, 156-159.
29. Shcherbakov, A.S., Luna Castellanos, A. and Sanchez Lucero, D. (2009) Characterization of the Beam Shaper and Fourier Transform System in a Prototype of the Acousto-Optical Spectrometer for Mexican Large Millimeter Telescope. Proceedings of SPIE, 7386, 1-12.

30. Shcherbakov, A.S., Sanchez Lucero, D., Luna Castellanos, A. and Maximov, Je. (2010) Some Peculiarities of Designing the Optical Scheme of Tellurium Dioxide Crystalline Cell Based Acousto-Optical Spectrometer for the Mexican Large Millimeter Telescope. Proceedings of SPIE, 7598, 1-10.

31. Herrera Martínez, G., Luna Castellanos, A., Carrasco Bazúa, L. and Shcherbakov, A.S. (2011) Testing the Bragg Cell of an Acousto-Optical Spectrometer for Radio Astronomy. Mexican Review of Astromomy& Astrophysics, Conference Series, 40, 305.

32. Shcherbakov, A.S., Sanchez Lucero, D. and Luna Castellanos, A. (2011) Global Characterization of an Advanced Prototype of a Multi-Channel Acousto-Optical Spectrometer for the Mexican Large Millimeter Telescope. Proceedings of SPIE, 7934, 1-12.

33. Shcherbakov, A.S., Sanchez Lucero, D., Luna Castellanos, A. and Maximov, Je. (2011) Characterizing the Polarization Features of a Multi-Prism Fused-Silica Beam Expander for a Wide-Aperture Acousto-optic Applications. Proceedings of SPIE, 7934, 1-11.

34. Korablev, O., Fedorova, A., Bertaux, J.-L., Stepanov, A.V., Kiselev, A., Kalinnikov, Yu.K. A., Titov, Yu., Montmessin, F., Dubois, J.P., Villard, E., Sarago, V., Belyaev, D., Reberac, A. and Neefs, E. (2012) SPICAV IR Acousto-Optic Spectrometer Experiment on Venus Express. Planetary and Space Science, 65, 38-57.http://dx.doi.10.1016/j.pss.2012.01.002

35. Helmich, F.P. (2011) Herschel HIFI—The Heterodyne Instrument for the Far-Infrared. European Astronomical Society Publications Series, 52, 15-20.Http://dx.doi.10.1051/eas/1152003

36. Born, M. and Wolf, E. (1999) Principles of Optics. 7th Edition, Cambridge University, Cambridge.

37. Goodman, J.W. (2005) Introduction to Fourier Optics. 3rd Edition, Roberts & Co., Greenwood Village.

38. Iizuka, K. (2002) Elements of Photonics. Vol. 1, John Wiley & Sons, New York.

39. Balakshy, V.I., Parygin, V.N. and Chirkov, L.E. (1985) Physical Principles of Acousto-Optics. Radio iSzyaz, Moscow.

40. Das, P.D. and DeCusatis, C.M. (1991) Acousto-Optic Signal Processing: Fundamentals & Applications. Artech House, Boston.

41. Klein, R.W. and Cook, B.D. (1967) A Unified Approach to Ultrasonic Light Diffraction. IEEE Transactions on Sonics &Ultrasonics, 14, 123-134. http://dx.doi.org/10.1109/T-SU.1967.29423

42. Dmitriev, V.G., Gurzadyan, G.G. and Nikogosyan, D.N. (1999) Handbook of Nonlinear Optical Crystals. 3rd Edition, Springer, Berlin. http://dx.doi.org/10.1007/978-3-540-46793-9

43. Blistanov, A.A. (2007) Crystals for Quantum and Nonlinear Optics. MISIS, Moscow.

44. Dirac, P.A.M. (1999) The Principles of Quantum Mechanics. 4th Edition, Oxford University Press, Oxford.

45. Shcherbakov, A.S., Bliznetsov, A.M., Castellanos, A.L. and Sánchez Lucero, D. (2010) Acousto-Optical Spectrum Analysis of Ultra-High Frequency Radio-Wave Analogue Signals With an Improved Resolution Exploiting the Collinear Acoustic Wave Heterodyning. Optik—International Journal for Light and Electron Optics, 121, 1497-1506.http://doi:1016/j.ijleo.2009.02.015
46. Shcherbakov, A.S., Sánchez Lucero, D., Luna Castellanos, A. and Belokurova, O.I. (2010) Direct Multi-Channel Optical Spectrum Analysis of Radio-Wave Signals Using Collinear Wave Heterodyning in Single Acousto-Optical Cell. Journal of Optics, 12, 045203.http://doi:10.1088/20408978/12/4/045203

CITATION

Shcherbakov, A. and Luna, A. (2014) Developing an Advanced Prototype of the Acousto-Optical Radio-Wave Spectrometer for Studying Star Formation in the Milky Way. *International Journal of Astronomy and Astrophysics*, **4**, 128-144. doi: 10.4236/ijaa.2014.41012.

Chapter 6

Solar Activity During the Rising Phase of Solar Cycle 24

Aradhna Sharma, S. R. Verma

Post Graduate Department of Physics, DBS (PG) College, Dehradun, India

ABSTRACT
Solar activity refers to any natural phenomenon occurring on the sun such as sunspots, solar flare and coronal mass ejection etc. Such phenomena have their roots deep inside the sun, where the dynamo mechanism operates and fluid motions occur in a turbulent way. It is mainly driven by the variability of the sun's magnetic field. The present paper studies the relation between various solar features during January 2009 to December 2011. A good correlation between various parameters indicates similar origin.

INTRODUCTION
The solar atmosphere continues to be one of the richest and most dynamic environments studied in modern astrophysics. Spanning many orders of magnitude in density and temperature, while linked to the complex system of magnetic field the sun displays a myriad of interesting phenomenon from sunspots in the photosphere to coronal mass

ejections—the most energetic events in the solar system. Like earth, the sun has seasons. More precisely, it has a cycle that lasts about 11 years. The number of sunspots rises and falls and rises again in about 11 years. This is due to the variability of solar magnetic field. The variability of the magnetic field has a strong influence on the dynamics of the outer layer of the sun and is registered by several solar parameters such as the sunspot number, the rate at which flux and coronal mass ejections occur, the flux of solar X-rays and radio waves.

The wealth of solar coronal phenomena called as solar activity should be viewed beyond their individual occurrences [1]. Coronal mass ejections are the most spectacular phenomenon of solar activity. CMEs occur in regions of closed magnetic fields that overlie magnetic inversion lines [2]. A study on CME is an important topic that is related directly to space environment [3]. The sunspot cycle is an important form of solar variability that indicates the extent of closed magnetic field structure on the sun, and hence is important to the study of the origin of coronal mass ejections. Based on the 110 Skylab CMEs, Hildner et al. [4] found the CME rate (R) to be correlated with the sunspot number and obtained the relation R = 0.96 + 0.084 N (based on 7 rotation)

They suggested that this relation is independent of the phase of the solar activity cycle and predicted a rate of 3.2 per day for solar maximum phase.

Webb & Howard [5] studied CMEs from 1973 to 1989 concluding that CME occurrence frequency tends to follow the solar activity cycle in both amplitude and phase. Gopalswamy et al. (2009) [6] have also studied CME occurrence in relation to sunspot number and found that the correlation between them is quite weak during the maximum phase period of solar cycle as compared to that in both ascending as well as descending phase. Researchers have studied the solar cycle that ended in December 2008 which is known as solar cycle 23. This cycle was longer than normal. The present solar cycle 24 started in December 2008 and is expected to have a shorter time period. In this paper we have studied the relation between various solar features during January 2009 to December 2011 for this cycle.

SOURCES OF THE DATA

Data were obtained from the SOHO-LASCO CME catalogue http://cdaw.nasa.gov/cme_lis/index.html,from the NOAA websites ftp://ftp.ngdc.noaa.gov/STP/SOLAR_DATA/SUNSPOT_NUMBERS/INTERN ATIONAL; ftp://ftp.ngdc.noaa.gov/STP/SOLAR_DATA/Flux/penticon_ observed,and ftp://ftp.ngdc.noaa.gov/STP/SOLAR_ DATA/Satellite_ enviornment/XRay_BGND/GOESBGND.o6 for coronal mass ejection frequency, for sunspot number,for 2800 MHz solar radio emission,and for solar X-ray background, in that order.

DATA ANALYSIS

Figure 1 shows the variation of different solar parameters during January 2009 to December 2011. The figure shows that the sun is very quiet with less sunspots and solar activity in the beginning of solar cycle 24.

Figure 1(a) shows the monthly occurrence frequency (R_w) of the coronal mass ejections (CME) given in SOHOLASCO catalogue. The lowest frequency has been 46 in August 2009. The frequency increases in the year 2010 to a maximum of in the month of November. In 2011 CME frequency increases further, reaching a maximum of 227 in the month of October.

Figure 1(b) shows a similar plot for maximum value of sunspot numbers ($R_z(max)$) during the same period. In August 2009, the sunspot number was zero. Like CME, sunspot number was much higher in the year 2011.
In **Figure 1**(c) a graph for 2800 MHz solar radio emission F-10 (units $10^{-22} \cdot Js^{-1} \cdot m^{-2} \cdot Hz^{-1}$) is plotted. This plot is very much similar to that of sunspot number.

Figure 1(d) shows a similar plot for the X-ray background ($W \cdot m^{-2}$). The X-ray data are available only up to February 2011. In the data table, low values are mentioned < 1.0. These have been set as 0.50. The matching between (a) & (d) is good.

For detailed analysis linear plots have also been plotted. **Figure 2" target="_self"> Figure 2**shows the linear plot for monthly occurrence frequency of CME (R_w) and sunspot no. (R_z(max)). The mismatch in peak occurrence is obvious. The correlation coefficient was found to be 0.906.

Figure 3 shows the linear plot for R_w and solar flux. The correlation between R_w and solar flux is 0.813009.

Figure 4 is the linear plot for R_w and intensity. The correlation between R_w and intensity is 0.325.

Figure 5 shows the linear plot between R_z(max) and solar flux. The correlation coefficient between R_z and solar flux is 0.88.

CONCLUSIONS

1) Solar cycle 24 has initially displayed much less activity;
2) CME occurrence frequency shows almost similar variational pattern with other forms of solar activity. This indicates similar origins, probably due to similar magnetic configuration affecting all parameter simultaneously;
3) R_w & R_z show very high & positive correlation. Kane 2011 [7] has found similar result for earlier cycle. But on comparing the plots of 1(a) & 1(b) we see that CME activity and sunspot cycle do not match exactly. It may be due to the fact that CMEs originate not only from sunspot regions but also from non-sunspot regions. Ramesh & Rohini [8] and Ramesh [9] have shown that CME frequency is better related with sunspot area than with sunspot numbers. However, Kane[10] has mentioned that sunspot areas and sunspot numbers are very highly correlated. Therefore, sunspot number, sunspot group number and sunspot area could be used as good proxies for each other;
4) CME occurrence frequency and solar flux show positive and high correlation with a correlation coefficient of 0.8;
5) R_w and X-rays are somewhat correlated with the correlation coefficient of 0.325. The low correlation may be due to the fact that data in some cases have been <1.0 and to which the GOES instrument is not sensitive;
6) R_z and solar flux show very high correlation. A good correlation between various parameters indicates similar origin.

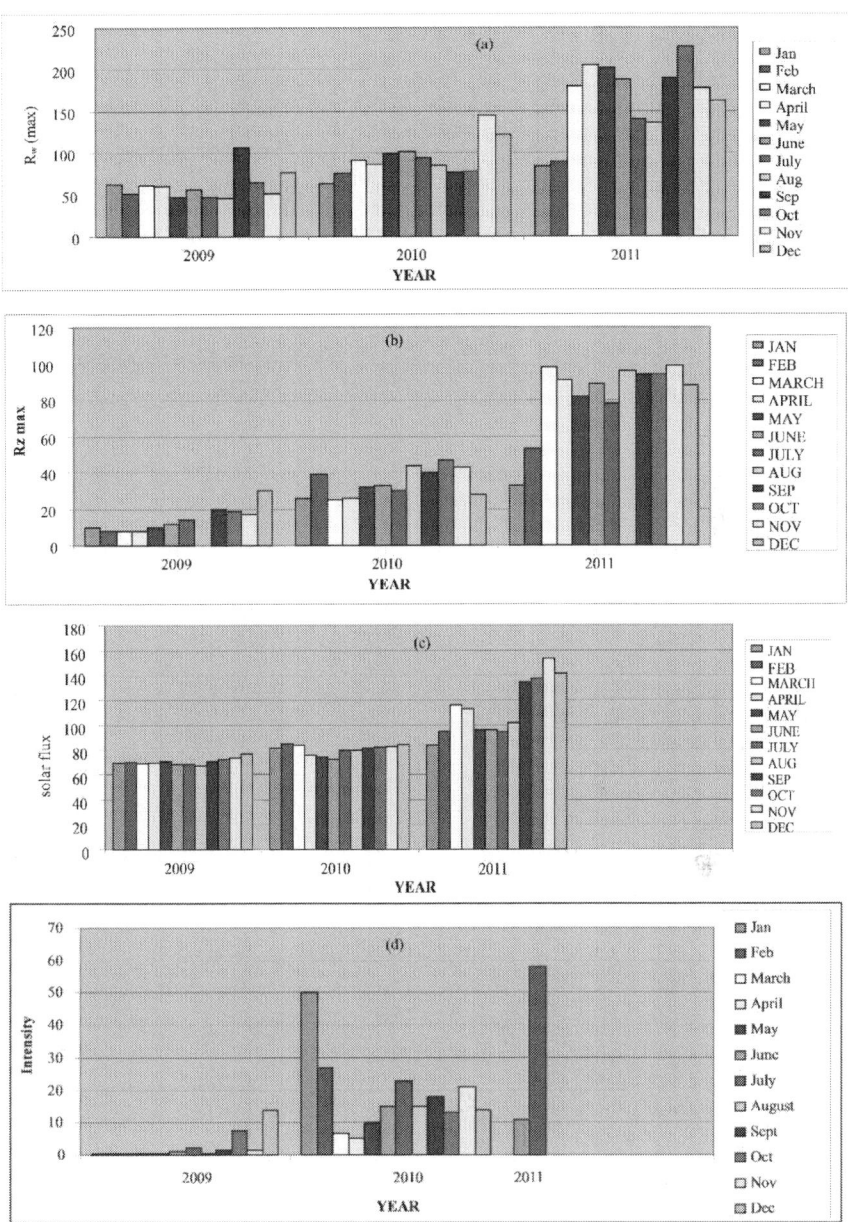

Figure 1. Plots of monthly values during Januray 2009-December 2011 for (a) CME occurrence frequency; (b) sunspot no. R_z (max); (c) 2800 MHz solar radio emission F-10 ($Js^{-1} \cdot m^{-2} \cdot Hz^{-1}$); (d) X-ray background (1 - 8 Å units $W \cdot m^{-2}$).

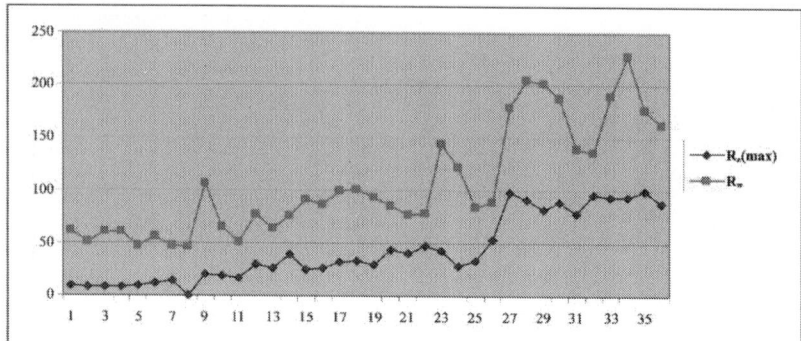

Figure 2. The linear plot for sunspot number R_z and CME occurrence frequency (R_w).

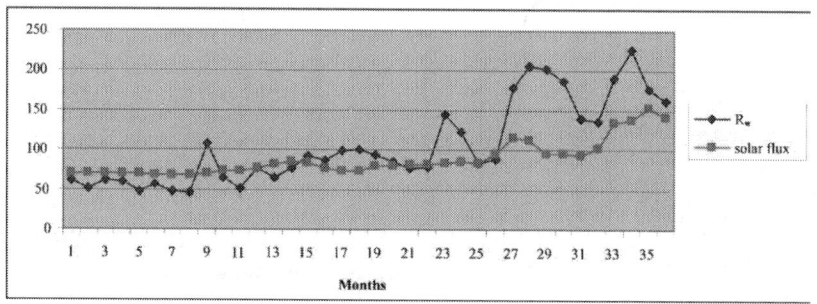

Figure 3. The linear plot for CME occurence frequency R_w & solar flux.

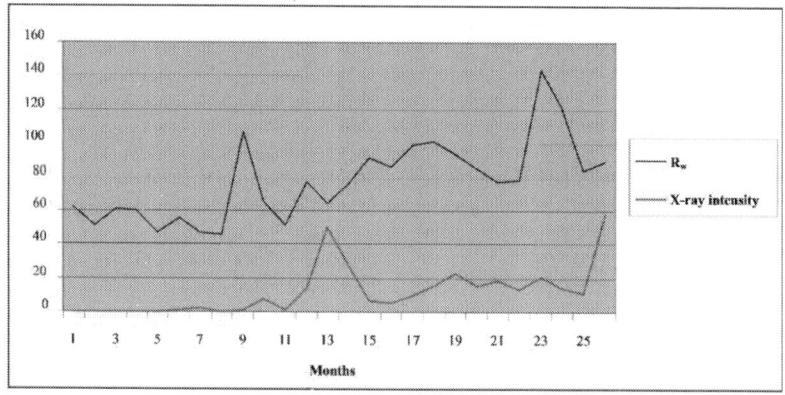

Figure 4. The linear plot for CME occurrence frequency R_w & intensity.

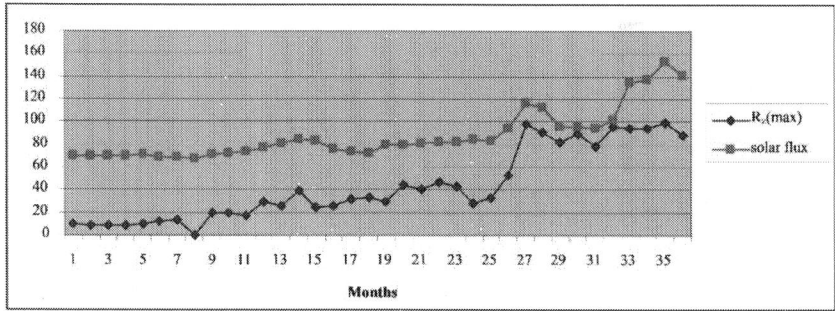

Figure 5. The linear plot for sunspot no. R_z and solar flux.

The study of coronal mass ejections and their relationship to other forms of solar activity provides an important link in the chain of evidence connecting all solar activity to its ultimate physical cause, the structure and evolution of the solar magnetic field. Understanding the solar magnetic fields has become highly important in the present scenario. We have to develop realistic models of the flares and CMEs because they are the main drivers for the space weather disturbances that strongly affect our high-tech life. During their propagation in the solar system, CMEs may frequently interact with the earth, producing a series of impacts on the terrestrial environment and human high-tech activities. Despite years of study we still don't understand key aspects of CMEs; specifically, how are they initiated in the solar corona, and how they evolve to produce the signatures that are measured with the interplanetary spacecraft. Watching the Sun continuously and analyzing the solar data constantly would make predictions of such events. Such quantitative relationships will be important for modeling studies and for space weather predictions.

ACKNOWLEDGEMENTS
We thank to various world data centers for providing the data.

REFERENCES

1. B. C. Low, "Coronal Mass Ejections, Magnetic Flux Rope and Solar Magnetism," Journal of Geophysical Research, Vol. 106, No. A11, 2001, pp. 25, 141-25163.
2. J. T. Gosling, "Coronal Mass Ejections and Magnetic Flux Rope in Interplanetary Space," AGU Monograph SeRies, Vol. 58, 1990, pp. 343-364.
3. N. J. Fox, M. Peredo and B. J. Thompson, "Cradle to Grave Tracking of the January 6-11, 1997, Sun-Earth Connection Event," Geophysical Research Letters, Vol. 25, No. 14, 1998, pp. 2461-2464.
4. E. Hildner, et al. "Frequency of Coronal Transients and Solar Activity," Solar Physics, Vol. 48, No. 1, 1976, pp. 127-135.
5. D. F. Webb and R. A. Howard, "The Solar Cycle Variation of Coronal Mass Ejections and the Solar Wind Mass Flux," Journal of Geophysical Research, Vol. 99, No. A3, 2012, pp. 4201-4220.
6. N. Gopalswamy, et al., "Magnetic Coupling between the Interior and Atmosphere of the Sun," In: S. S. Hasan and R. J. Rutten, Eds., Astrophysics and Space Science Proceedings, Springer-Verlag, Berlin, 2010, pp. 289-307.
7. R. P. Kane, "Solar Activity during Sunspot Minimum," Indian Journal of Radio & Space Physics, Vol. 40, No. 1, 2011, pp. 7-10.
8. K. B. Ramesh and V. S. Rohini, "1-8 Angstrom Background X-Ray Emission and the Associated Indicators of Photospheric Magnetic Activity," The Astrophysical Journal Letters (USA), Vol. 686, No. 1, 2008, pp. L41-L44.
9. K. B. Ramesh, "Coronal Mass Ejections and SunspotsSolar Cycle Perspective," The Astrophysical Journal Letters (USA), Vol. 712, No. 1, 2010, pp. L77-L80.
10. R. P. Kane, "Similarities and Dissimilarities between the Variations of CME and Other Solar Parameters at Different Heliographic Latitudes any Time Scale," Solar Physics, Vol. 248, No. 1, 2008, pp. 177-190.

CITATION

A. Sharma and S. Verma, "Solar Activity during the Rising Phase of Solar Cycle 24," *International Journal of Astronomy and Astrophysics*, Vol. 3 No. 3, 2013, pp. 212-216. doi: 10.4236/ijaa.2013.33025.

Chapter 1

Study Chromaticity of Solar Spectrum

Nagendra Nath Mondal[1,2]

[1]Department of Physics, Batanagar Institute of Engineering, Management and Science (BIEMS) (Techno India Group), Batanagar, Kolkata 79, West Bengal, India
[2]Department of Physics, Indian Institute of Engineering, Management and Science (IIEST), Shibpur, Howrah 711 103, West Bengal, India

ABSTRACT

The chromaticity of solar spectrum is studied with the help of a Solar Spectrum Monitor system that can detect individual color in the spectrum. Recent observations done by the detector on the Solar Radiation and the Erath Milieu Experiment suggest that the sun's visible spectral irradiance changes from May 2009 to September 2012. The data of Earth's coordinates and environment have been taken since April 2005 after the devastating Tsunami (December, 2004) of India. The bizarre data of zenith angle, azimuth angle, and temperature of the Earth's atmosphere show their changes of maxima and minima epoch to epoch. The data of solar spectrum monitor have been taking since 2009 and significant transformations of colored ratios $\Delta R_{B/W}$ and $\Delta R_{R/W}$ per hour are observed among the regions of each solstice between 2009 and 2012. The author advocates that the abrupt vagaries of the Earth's movements may cause devastating tsunamis, earthquake, volcanic eruption, cyclones and tornadoes in addition of anomalous changes of solar spectral irradiation, humidity, temperature and pressure; those effects may spoil ecological balance and extinct some living species from the soil of the earth.

INTRODUCTION

The most challenged questions among the nineteenth century scientists were what makes the sun shine and what are the sources of huge amount of energy necessary to support lives on Earth? In the middle of 20th century we come to know the exact answers of those questions when we discovered nuclear fission and fusion and the role of solar irradiation on our lives [1] . The Sun is a few billion years old and its ultimate fate is a White Dwarf like many other stars in the universe. Hence, are the solar luminosity, chromaticity and the trajectory of the Earth quiet perpetual? The objective of this study is to find out specific answers of these crucial questions and their impacts on our lives. As the debate wraths over global warming, nearly every scientist will sustain the fact that the Earth has gotten warmer in the last 4 - 3 decades. What continues to be debated, and rightfully so, is the reason for the warming trend. Since about 1950, the Earth's global surface temperature has risen by just more than 0.6 degrees Celsius or just over 1 degree Fahrenheit, and the decline of Arctic Sea Ice nearly 32% in 2007 measured in compare to 1979, from 1870 to 1992 the average sea level rise has been 1.7 mm. Before 1950 CO_2 level was almost steady (180 - 300 ppm), but level is drastically increased and in 2007 it is almost 400 ppm [2] . A lot of researches have been carried out in the astronomy and astrophysics. So far we are studying the chromaticity of solar spectrum and presenting results for the first time. Experimental observations of solar luminosity (L_S), angular momentum and atmospheric temperature of the Earth may help us to resolve some of the fundamental problems on global environments and that can be found elsewhere [3] . In the following sections experimental procedure, data taking, data analysis, results and discussions will be illustrated.

Origin of Solar Spectrum

(A) Source of Light The light is a visible range of an electromagnetic (EM) wave that has been producing by fusion reaction in the core of Sun since its birth. Enrico Fermi discovered the fusion reaction and proton-proton (p-p) chain reaction is one the most dominant source of continuous irradiation of EM wave among the other sequences which is shown in Figure 1.

From this single reaction cycle we notice that positron (e^+) an antiparticle of electron (e^-), γ-rays and huge amount of energy (1.44 MeV) are generated. After a complete cycle two products of ^1_1H are attained and make the successive reaction again. It is the source of Green energy and that is better than fission reaction in terms of their mass-energy ratio.

(B) Characteristics of EM Waves Positron is another source of EM wave that wave is generated over e^+e^- annihilation. When EM wave is penetrating a huge continuum (a few km) of photosphere different types of scatterings, e.g., Thomson scattering, Rayleigh scattering, Photoelectric effect, Compton scattering are taken place depending upon the various circumstances of the photosphere. Visible spectrum originated mostly from the atomic transitions of H (λ = 656 nm (red), λ = 486 nm (green)) and He atoms (λ = 588 nm (yellow), λ = 447 nm (violet)). The visible range of EM wave (200 nm - 900 nm) forms a solar spectrum which is depicted in Figure 2. It follows the Planck's radiation formula. Rayleigh scattering is the most dominant part in this continuum because of maximum numbers of neutral H is in the ground state. Hence significant information about the physical conditions of the Sun can be achieved by studying the chromaticity in different moments of the year.

Major thrust of this exertion is to study the VIBGYOR region by the Solar Spectrum Monitor (SSM) detector system that is described in the following subsequent sections.

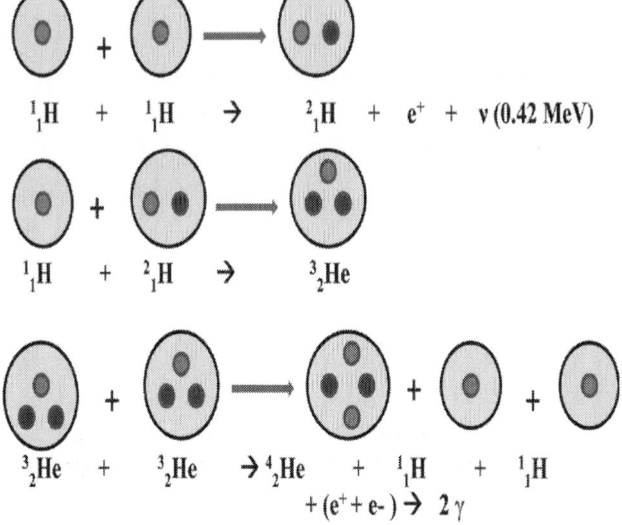

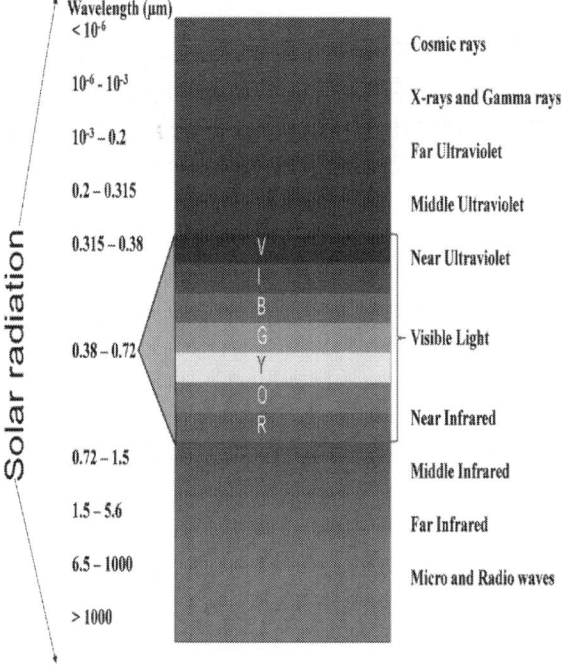

Figure 1. Proton-Proton chain reaction, a source of solar spectrum and energy.

Figure 2. Characteristics of solar spectrum.

THEORY

In order to determine the earth's Zenith angle (θ) and luminous currents by the SSM detector system that consists of colored plates (Red ® R, Blue ® B and White ® W) coupled with photodiodes through optical fiber and a measurement table (see sec. IV), and θ can be determined by the following equation.

$$\theta = \tan^{-1}\left(\frac{S}{L}\right)$$

(1)

where, S and L respectively are the shadow length and length of the spike that is placed perpendicular on the table. Solar luminous current of a colorless plate (W) can be determined by the following equation:

$$I_{W,S} = S_{G,pd}\left(\frac{L_{W,S}}{4\pi R^2} \times \Omega_G\right) \times U_{\lambda,W}$$

(2)

where $I_{W,S}$ and $L_{W,S}$ respectively are current intensities due to W plate and solar luminosity, R is the distance between the Earth and the Sun. The luminous sensitivity of epoxy glass with photodiode is $S_{G,pd}$ which is assumed to be independent of the specific color band. Ω_G is the area of solid angle exposed by the solar spectrum, i.e.,

$$\Omega_G = \int_0^{\pi/2} \sin\theta d\theta \int_{\pi/2}^{3\pi/2} \cos\varphi d\varphi$$

(3)

and absorbed photons follow the Planck's distribution which is given by

$$U_{\lambda,w} = \frac{2\pi hc^2}{\lambda_w^5\left(e^{hc/\lambda_w KT} - 1\right)}$$

(4)

where h is Planck's constant, c is the speed of light, λ_w is the wave length. Similarly $I_{B,S}$ and $I_{R,S}$ can be determined respectively from the following equations:

$$I_{B,S} = S_{G,pd}\left(\frac{L_{B,S}}{4\pi R^2} \times \Omega_G\right) \times U_{\lambda,w}$$

(5)

$$I_{R,S} = S_{G,pd}\left(\frac{L_{R,S}}{4\pi R^2} \times \Omega_G\right) \times U_{\lambda,w}$$

(6)

The strategy of this measurement is to find out any change of luminosity of color plates that will envisage transformations in the reaction mechanisms inside the core of the Sun. When extraterrestrial photon enters into the earth's atmosphere its intensity can be reduced by the absorption of molecules at different atmospheric layers. It is impossible to measure accurate $L_{W,S}$, $L_{B,S}$ and $L_{R,S}$ with greater precision from the soil of the earth than in space. Therefore color luminous ratio with respect to whole spectrum of W plate will provide better information and understanding about the present status of the solar spectrum as well as the Sun. The ratio $R_{IB/IW}$ and $R_{IR/IW}$ can be derived using (2), (5) and (6) which are given below:

$$R_{IB/IW} = \frac{L_{B,S}}{L_{W,S}} = \frac{I_{B,S}}{I_{W,S}}$$

(7)

and

$$R_{IR/IW} = \frac{L_{R,S}}{L_{W,S}} = \frac{I_{R,S}}{I_{W,S}}$$

(8)

Experimentally we have accumulated huge data over the years and analyzed in order to achieve $R_{IB/W}$ and $R_{IR/W}$.

EXPERIMENTAL PROCEDURES

(A) Detector Development In Figure 3 a simple and inexpensive SSM detector is depicted which consists of Silicon Pin Photodiodes (model: BPW 34), colored B, R and W epoxy glasses, optical fibers and a Micro-ammeter. Details of the SSM detector development and experimental procedure can be found elsewhere [3] . In order to study the chromaticity of the solar radiation spectrum data was taken throughout the year and year after year in every weekend and almost in every national holiday between 7:00 to 17:00. Colored luminous absorption currents produced by the solar irradiation exposed directly on to the plate surfaces were recorded by the micro-ammeter manually in each coordinates of the Earth about an hour interval.

(B) Measurement of Orbital and Rotational angles The orbital and the rotational motions of the Earth respectively are about the Sun and its own axis. A round table scaled with a protractor in four quadrants by marking the East to West and North to South lines. An iron spike of length 20 cm is fixed at the center of the table for the determination of the θ and the ϕ angles respectively from the shadow length and from the protractor scale. The measurement table is calibrated with the help of a Compass.

In Figure 4 θ, ϕ, equator (ABD), prime meridian (NGJS) etc. are illustrated. In each θ and ϕ, T was recorded by a Mercury thermometer. Experiments were performed in Kalyani, a city of West Bengal, India about 22°39'N from the Equator, 88°27'E from the Prime meridian and about 11 m above the Sea level. Equator is passing through the middle of India. Hence Southern parts of India are warmer (e.g. Tamil Nadu, Kerala) than Northern parts (e.g. Jammu & Kashmir and Himachal Pradesh which are full of Snow in winter).

Hence, we determine θ and ϕ respectively from the recorded shadow length and the position on the scale.

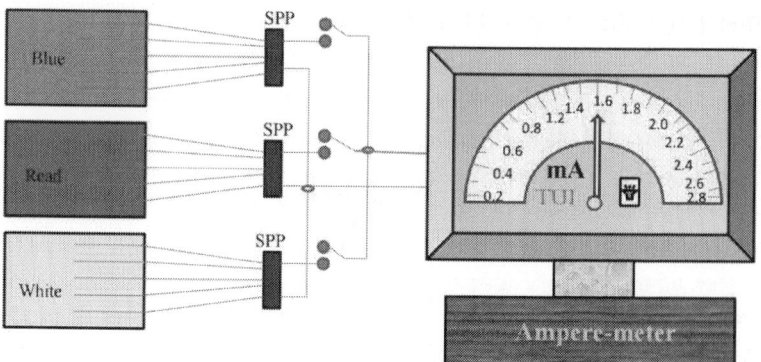

Figure 3. SSM detector is exposed directly by the solar spectrum.

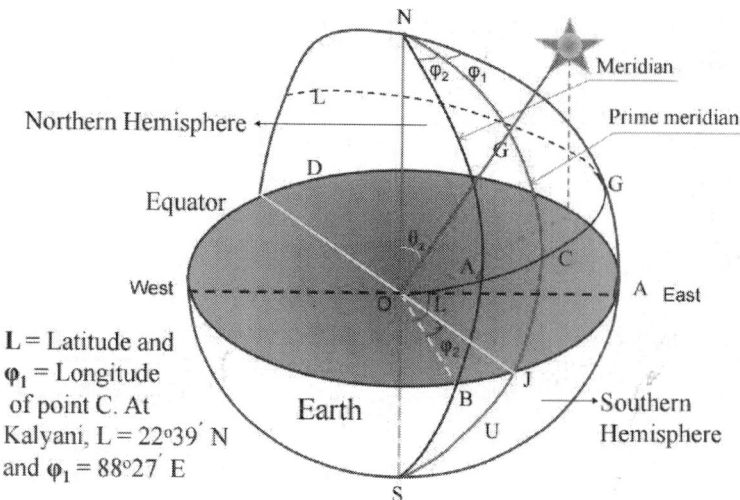

Figure 4. A Global view of the Earth with coordinates.

RESULTS AND DISCUSSIONS

(A) Zenith angle (ϑ):
A typical spectrum of different measurements carried out between May 2009 and September 2012 is depicted in Figures 5(a)-(f). The spectrum (d) shows the distribution of shadow length S, which can be used to determine θ distribution from (1). Global definition is exactly reflected in

this measurement where we have noticed that S is minimum in summer, i.e., θ is minimum due to the equator as the Sun is on the above of our heads (~12:00). In winter S goes to maximum as the earth's rotational axis tilted towards the north, hence S becomes longer i.e., range of θ gets narrower than that of summer. The maximum and the minimum of each spectrum (d & e) respectively show the data of around December and June.

In Figure 5(e) the spectrum of φ is shown. During the summer (March-August) and winter (SeptemberFebruary) solstices θ and φ attain to minimum and maximum. The summary of θ at different times (7:00-17:00) in solstices are shown in Figure 6 where maxima and minima of θ clearly infer the discrepancies among the years. Each peak between summer solstices and winter solstices indicate respectively the perihelion and equatorial motions. It concludes that equatorial motion is slower than the perihelion motion since diverse time periods are relating in these two regions. The summary of the measurement of θ among summer solstices and winter solstices are presented distinctly from where the drastic change of θ before and after the line of Tsunami is evidently described.

(B) Azimuth angle (φ):
Similarly φ is estimated and the summary of a typical measurement at ~9:00 is depicted in Figure 7. The significant discrepancies of φ among the years are deliberating slants of major axis of rotation in every year.
Especially before and after the Tsunami line the change of φ is remarkable and that brings a huge devastation. The dateline of Tsunami in March 2011 is depicted in the spectrum from where the difference of φ is estimated to be ~6°.

(C) Temperature (T):
The recorded data of temperature is plotted in Figure 5(f). The period of lower temperature measurement in winter solstice is shorter than that of higher temperature in summer solstice. The peak positions of each spectrum (d, e, f) are assumed to be symmetric and rotational and orbital axis are not changing. But we have found inconsistencies when we excavate the data profoundly. The overturned spectra of temperature indicate that ray of light from the Sun fall obliquely on to the surface of the Earth; hence temperatures dwindle in winter sharply. The summary of

the measurement of T is depicted in Figure 8 where difference of temperatures before and after the Tsunami is estimated. Results bring the same phenomena which we have concluded previously in the case of θ and φ respectively.

Environmental pollution is another factor that can change T enormously. Maximum T can be observed at noon and minimum respectively in the morning and in the afternoon. In the summer air pollution is minimum and T goes to maximum while in the winter T goes to minimum.

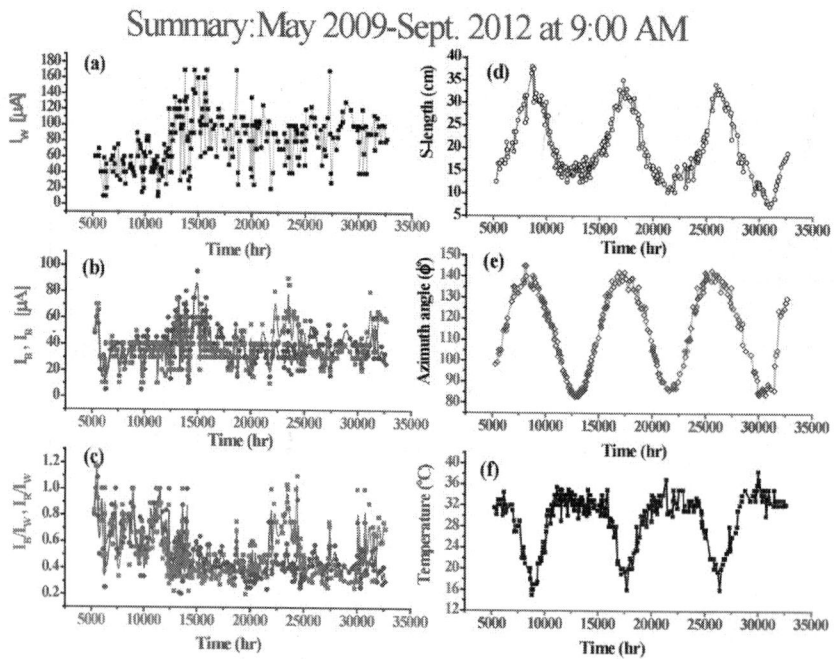

Figure 5. A typical summaries of astrophysical ((a)-(c)) and astronomical ((d)-(f)) data of the measurement.

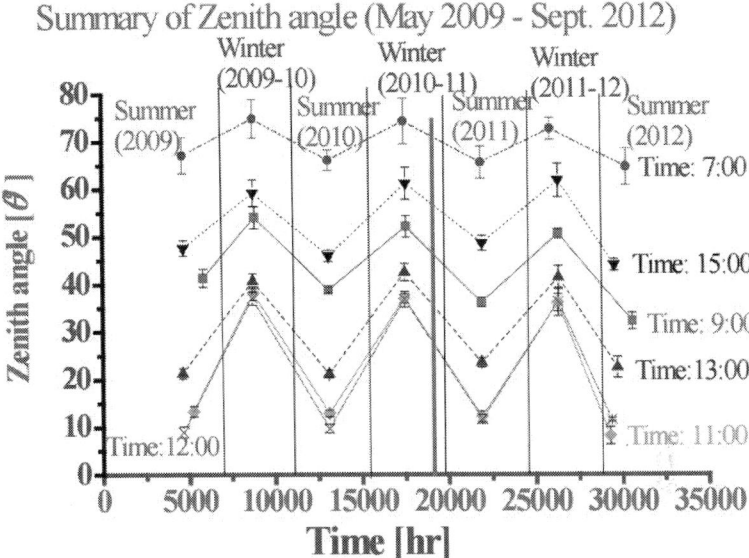

Figure 6. Spectra show the deviation of θ at different solstice regions which are indicated by the columns.

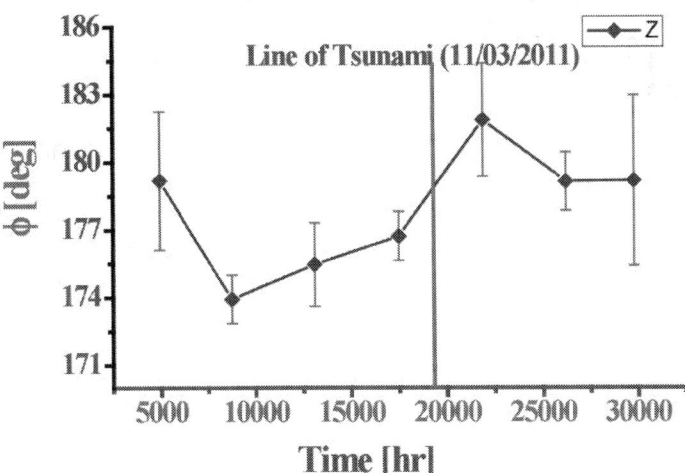

Figure 7. Spectrum shows the distinct variation of φ before and after the Tsunami.

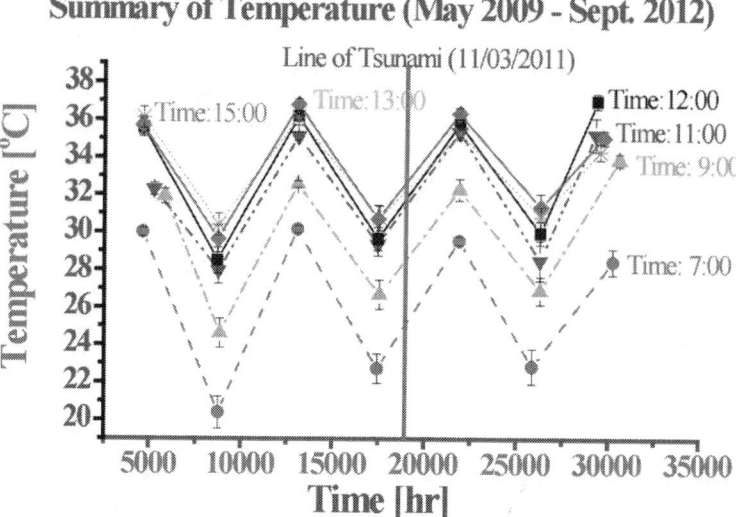

Figure 8. Variations of temperature are offered.

(D) Chromaticity:

In order to study the chromaticity of a solar spectrum the SSM detector system is used. The distributions of current produced by W, B and R plates are depicted in Figures 5(a)-(c). In (a) currents of W, (b) currents of B & R (overlapped) and (c) ratios of I_R/I_W and I_B/I_W are presented. Those spectra don't have any peak as can be seen at right column. But the advents of maximum in summer and minimum in winter for a discrete period are visible due to the Earth's two periodic motions which are shown in the right column of Figures 5 (d)-(f).

The intensity distribution of red light is dominated over the blue light and that is visible in the overlapping spectra. In the data analysis first of all we calculate average current after subtracting the background of each color plate considering the periods of summer and the winter solstices. Secondly we obtain ratios between I_B & I_W (Figure 9) and I_R & I_W to determine their differences and time period between same solstices among the years. Thirdly we divide these ratios by the difference of time interval (unit is assigned to be ppm) and plot them with respect to the mean period of the solstices.

One of the typical spectra of $\Delta R_{IB/IW}$ is shown in Figure 10 that was analyzed from the data taken at ~9:00. Significant variations of colored ratios per hour between the similar solstices are observed. The rate of decrease of $\Delta R(I_B/I_W)$ from 2009 to 2010 in summer solstice is (7.46 ± 3.20) ppm and from 2010 to 2011 is (10.11 ± 0.32) ppm. We also determined the rate of decrease of $\Delta R(I_B/I_W)$ from 2009 to 2010 in winter solstice is (20.61 ± 2.21) ppm. Similar phenomena also can be observed in the rate of decrease of $\Delta R(I_R/I_W)$ which is tabulated in the 7th column of Table3 The error bars of each point infer the statistical error. Bigger error bar represents smaller number of statistics over the six months period than the other points. Some times in summer (March-April-May), rainy seasons (June-July) and early autumn (August) the sky has various conditions: e.g., deep cloudy, cloudy, white nomadic clouds, blue sky, and bright blue sky just after the heavy shower. Hence variations of intensities observed more in the summer than that of winter solstices.

The change of colored ratios implies the decaying nature of intensities in the solar spectrum and abrupt flora of it can be noticed (see the ring) before and after the Tsunami. Similar data analysis is done for the I_R/I_W which is tabulated in Table1 Results attribute the decline nature of I_R/I_W. In this case we can also estimate the rate of change of ratios between the same solstices which are arranged in Table2 The differences between odd-th and even-th numbers respectively refer to the periods of summer and winter solstices. We have perceived the sudden change of R before and after the Tsunami line here too (4th row). The rates of changes of astronomical and astrophysical parameters estimated in this study are tabulated in Table3 From these observations we confer that the unusual motions of the Earth not only bring vigorous destruction but also harm our environment equally.

Any change of intensities of L_S alters the humidity that may reflect in the monsoon of a tropical area. The abrupt variation of monsoon causes on our environment in diverse ways. For example excess rainfall in the SouthEast Asia brings huge floods that spreads epidemic in the post era. On the other hand seldom rainfalls bring drought in many parts of the North-West regions, results scarcity of food and drinking water. Moreover the oxygen production by Photosynthesis depends on the colors and its intensities as well. The entire creature must be suffered from the insufficient oxygen production in nature in various ways especially in the respiratory systems. Hence these studies create exclusive spaces for multidisciplinary researches in an outstanding domain.

What a big threat it is in our lives!? Many interesting challenged questions result from this measurement and some of those are discussed sequentially. More advanced detector system is required in order to find out the discrepancies among the astrophysical parameters in the solar spectrum. Our future work is to develop an advanced detector system for the search of celestial Positronium, the reason of global warming and related ecological problems.

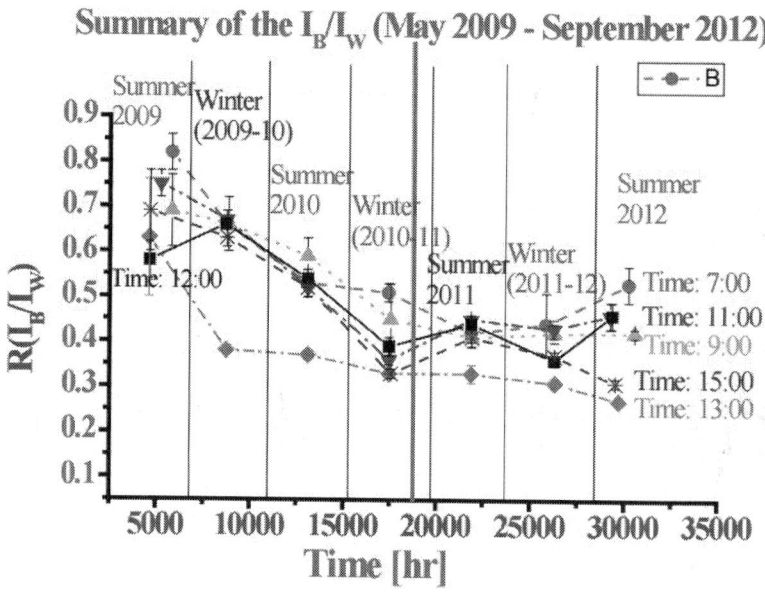

Figure 9. Variations of I_B/I_W over the years [hr] at different moments of days are shown.

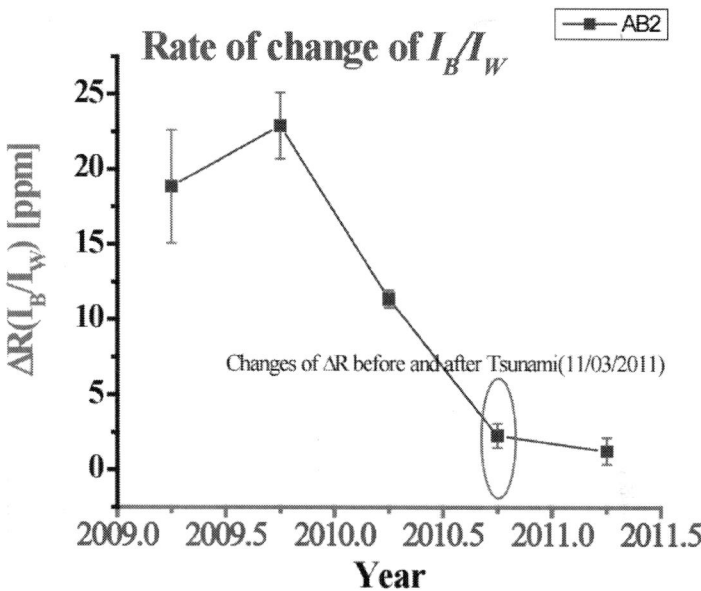

Figure 10. Spectra show the rate of changes of current ratios. Error bars indicates statistical error.

Table 1. Summary of I_R/I_W

Solstices	Average Time (hr)	Average I_R/I_W
Summer-2009	5137.33	0.68 ± 0.01
Winter-2009-10	8758.17	0.59 ± 0.02
Summer-2010	13096.92	0.56 ± 0.02
Winter-2010-11	17487.17	0.44 ± 0.03
Summer-2011	21885.08	0.58 ± 0.02
Winter-2011-12	26219.33	0.49 ± 0.02
Summer-2012	29779.67	0.33 ± 0.07

Table 2. Summary of rate of change of (I_R/I_W).

Differences of Solstices	ΔT [hr]	$\Delta I_R/I_W$)	Rate $\Delta_{IR/IW}$
1st - 3rd	7959.59	0.12 ± 0.06	15.1×10^{-6}
2nd - 4th	8729.00	0.15 ± 0.01	17.2×10^{-6}
3rd - 5th	8788.16	0.02 ± 0.01	2.27×10^{-6}
4th - 6th	8732.16	0.05 ± 0.01	5.72×10^{-6}
5th - 7th	7894.59	0.25 ± 0.05	31.7×10^{-6}

Table 3. Summary of rate of change of astronomical & astrophysical parameters.

Differences of Solstices	$\Delta R_s [\times 10^{-6}]$	$\Delta R_\infty [\times 10^{-5}]$	$\Delta R_r [\times 10^{-6}]$	$\Delta R_{pd} [\times 10^{-4}]$	$\Delta R_{IR/IW} [\times 10^{-6}]$	$\Delta R_{au/W} [\times 10$
Summer 1st - 3rd	10.80 ± 2.88	45.5 ± 14.80	9.96 ± 1.79	4.23 ± 1.06	18.84 ± 3.77	15.11 ± 7.5
3rd - 5th	6.14 ± 1.48	72.9 ± 7.51	4.32 ± 1.82	3.60 ± 0.04	11.38 ± 0.57	2.27 ± 0.5
5th - 7th	31.20 ± 3.74	26.9 ± 16.1	1.26 ± 0.29	5.56 ± 0.06	1.27 ± 0.89	31.70 ± 6.2
Winter 2nd - 4th	22.91 ± 2.78	33.60 ± 2.34	16.61 ± 2.29	4.73 ± 0.03	22.91 ± 2.29	17.23 ± 1.1
4th - 6th	9.27 ± 2.63	27.71 ± 2.29	4.58 ± 8.02	0.70 ± 0.23	2.29 ± 0.80	5.72 ± 1.1

CONCLUSION

We have developed a SSM detector system and measured the rate of change of color luminosities in terms of currents produced by color plates and determine the trajectories of the Earth's movement. This study may open a new domain of researches: 1) the Sun consists of many cells of different atoms and longevities. Hence sporadic rate of fusion reaction is possible to observe, 2) the huge magnetic field, ejection of solar flare, sun's spots, electron-positron plasma etc. are the sources of versatile luminosities in the solar spectrum, 3) Luminosities of individual color bands are changeable, which are observed in this study, 4) the alteration of chromatic intensities play crucial roles in ecological imbalance, especially the creation of oxygen in Photosynthesis, falling of solar energy, and those studies are going on, and 5) the change of intensity of color bands and color in the solar spectrum may excavate rare syndromes by congenital process or by revitalization, those can extinct many species of lives from the soil of Earth.

ACKNOWLEDGEMENTS

The author is very grateful to his family members for conducting this experiment in the house premises and help with financial support and cooperation.

REFERENCES

1. Bahcall, J.N. (1969) Neutrinos from the Sun. Scientific American, 221, 28-37.http://dx.doi.org/10.1038/scientificamerican0769-28
2. Lean, J.L. and DeLand, M.T. (2012) How Does the Sun's Spectrum Vary? Journal of Climate, 25, 2555-2560. http://dx.doi.org/10.1175/JCLI-D-11-00571.1
3. Mondal, N.N. (2014) Versatile Nature of the Earth Causes Disasters. J. Environ. Res. & Develop., 9, 1-8.

CITATION

Mondal, N. (2014) Study Chromaticity of Solar Spectrum. *International Journal of Astronomy and Astrophysics*, **4**, 510-518. doi: 10.4236/ijaa.2014.43047.

Chapter 8

On Some Statistical Characteristics of Radio-Rich CMES in the Solar Cycles 23 and 24

Joginder Sharma[a], Nishant Mittal[a,b], Udit Narain[a]

[a] Astrophysics Research Group, Meerut College, Meerut 250001, India
[b] Dept. of Physics, H.R. Institute of Technology, Ghaziabad 201003, India

ABSTRACT

In this paper we have presented the properties of radio-rich coronal mass ejections (Cmes), during the period 1997–2013. The CME event accompanied by the type II radio burst is referred to as radio-loud (RL), while the one lacking a type II burst is termed radio-quiet (RQ). These radio rich Cmes produce type II (1–14 MHz), i.e. decametric–hectometric or DH radio burst. It is found that the average width of all DH CMEs during the study period is 235° and 75% of the DH Cmes are halo Cmes in solar cycle 24. The DH CMEs linear speeds distribution is in the range 112–3387 km/s, with an average speed of 1043 km/s; the acceleration varies between 434 m/s^2 and −179 m/s^2. About 62% of the DH Cmes are decelerated. A CME associated with a type II burst and originating close to the center of the solar disk typically results in a shock at Earth in 2–3 days and hence can be used to predict shock arrival at Earth.

INTRODUCTION

Coronal mass ejections (CMEs) are the most energetic eruption in the solar system. Previous studies have established that many CMEs are associated with flares, and some of the physical properties of CMEs and flares are closely related; the flare, CME events which are associated with type II are said to be radio-loud, while the others without type II are named to be radio-quiet. Type II solar radio bursts have been studied by solar astronomers for more than 50 years. Wild and McReady (1950) first reported the observations of type II radio bursts from the dynamic spectra of solar radio bursts. The dynamic spectrum of a type II burst shows an emission band drifting from high to low frequency with a drift rate of <0.5 MHz/s, (Subramanian and Ebenezer, 2006). There is a general consensus that the type II radio bursts are a signature of shock wave propagating away from the Sun (Uchida, 1960, Nelson and Melrose, 1985 and Gopalswamy et al., 2005). It has been established that shocks driven by CMEs are responsible for interplanetary type II bursts observed at decametric–hectometric wavelength (Gopalswamy et al., 2001) and kilometric wavelengths (Cane et al., 1987). However, the drivers of shocks associated with the metric (or coronal) type II radio bursts are still controversial. CME-driven shocks as in IP type II bursts (Cliver et al., 1999,Gopalswamy et al., 2009 and Cho et al., 2011) and flare blast waves (Vrsnak et al., 1995,Khan and Aurass, 2002 and Magdalenic et al., 2008) are the two possibilities (Yashiro et al., 2014). DH and km type II radio bursts are recorded in the 1–14 MHz and 20 kHz – 1 MHz frequency ranges by RAD2 and RAD1 instruments respectively, constituting the Radio and Plasma Wave (WAVES) experiment (Bougeret et al., 1995) on board the wind spacecraft (Acuna et al., 1995).

In this paper we study the properties of DH-type II (1–14 MHz), i.e. radio-rich CME occurred during the period January 1997–May 2013 for the solar cycle 23 and 24. Earlier (Sharma et al., 2008 and Gopalswamy et al., 2005) have investigated the properties of radio-rich CMEs during the period January 1997–November 2006, with 367 events, and found that these DH CMEs are relatively faster and wider than the normal CMEs. The DH type II bursts provide an opportunity to remotely observe the formation and propagation of shocks in the solar corona and in interplanetary (IP) spaces.

DATA AND RESULT

The data (such as speed, angular width, acceleration and occurrence rate of CMEs) in this study are selected through the on line Coordinated Data Analysis Workshop (CDAW,http://cdaw.gsfc.nasa.gov/CME_list) database for CMEs observed by the Large Angle and Spectrometric coronagraph (LASCCO) onboard SOHO (Bruckner et al., 1995). During the solar cycle 23 and 24 in total more than 23000 CMEs have been observed till now by this instrument. Type II radio bursts were observed by Wind/WAVES spacecraft, (Bougeret et al., 1995) which is available on http://ssed.gsfc.nasa.gov/waves/burst. Since different observation use different density models for the calculation of the estimated shock speeds and different methods to determine the characteristics of the bursts, we have selected the events observed by Wind/WAVES spacecraft. A total of 638 events were observed by spacecraft from December 1994 to May 2013. Out of which 48 seem to be type IV radio bursts, and remaining 590 are type II burst. In our study we have considered 471 DH CMEs during the period 1997–2013, for the solar cycle 23 and 24. In solar cycle 24 (January 2009–June 2013) 95 events are taken.

Table 1 gives the annual distribution of average width, average speed and their width median, speed median along with standard deviation. Fig. 1 gives the distribution of speed during the study period but Fig. 2 shows the distribution of width of events of the same period and the variation between them is given in Fig. 4. It is found that the most of the events have speed greater than 800 km/s and width greater than 200°. It is also noted from Fig. 4 that DH CMEs having more width have greater speeds.

Table 1: Annual number, average speed, average width, median speed, median width and standard deviation of 471 DH CMEs.

Year	Number	Average width (°)	Median width (°)	Standard deviation of width	Average speed (km/s)	Median speed (km/s)	Standard deviation of speed
1997	10	199	118	132	590	428	372
1998	22	262	360	118	1038	1009	438
1999	17	251	360	123	1062	1025	568
2000	94	190	136	143	801	700	451
2001	56	234	271	134	1117	1008	588
2002	45	257	360	134	1211	1092	598
2003	34	225	226	128	1279	1101	583
2004	35	218	214	136	1067	925	594
2005	47	278	360	128	1295	1380	764
2006	12	161	67	148	853	829	554
2007	3	210	164	109	1106	995	185
2008	1	112	112	–	1103	1103	–
2009	–	–	–	–	–	–	–
2010	6	216	165	104	883	845	402
2011	37	275	360	111	1001	924	518
2012	43	300	360	106	1158	1138	509
2013	9	210	199	147	866	663	507

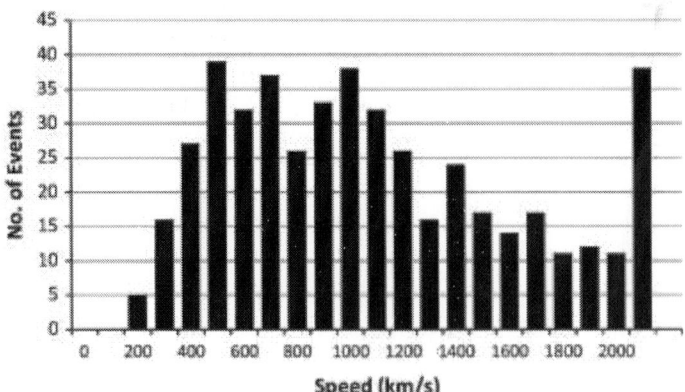

Figure 1: Histogram showing the speed distribution of 471 DH CMEs, during the period 1997–2013.

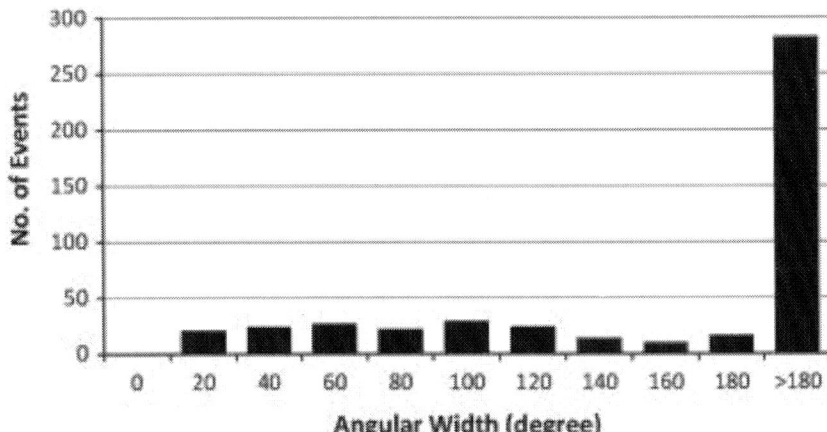

Figure 2: Histogram showing the angular width distribution of 471 DH CMEs during the period 1997–2013.

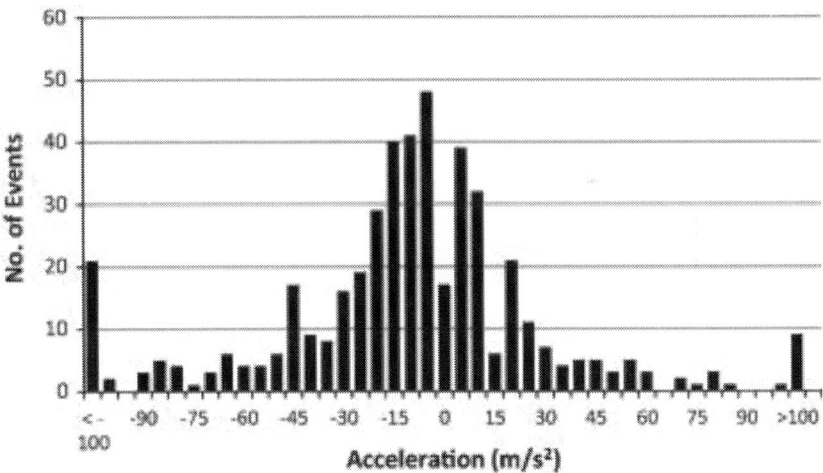

Figure 3: Histogram showing the acceleration distribution of 471 DH CMEs, during the period 1997–2013.

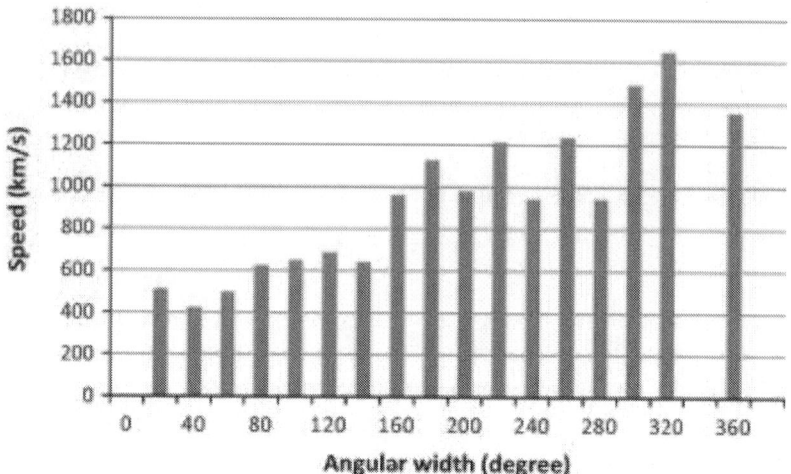

Figure 4: Histogram showing the distribution of angular width with the speed during the period 1997–2013.

Fig. 3 gives the distribution of acceleration, which shows that the most events are decelerated that is about 62%, but about 34% show positive acceleration and remaining move with little acceleration.

Fig. 5 gives the relation between the annual variations of angular width and speed. There is a remarkable change in the speed width relationship through the study period.

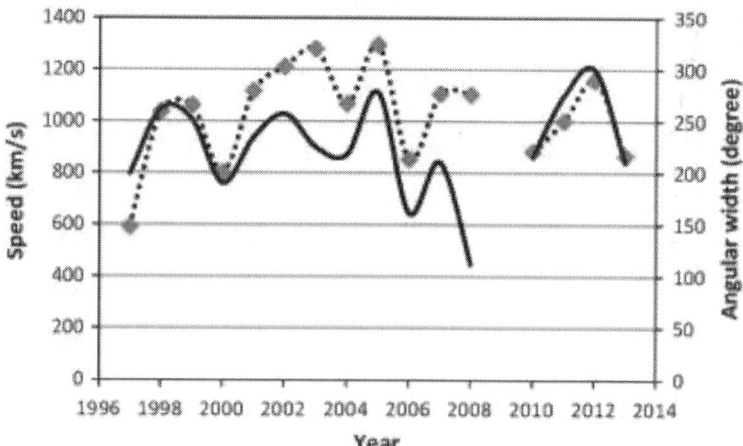

Figure 5: Annual variation of speed and angular width. Dotted line shows the speed distribution and angular width distribution is represented by solid line.

Fig. 6 shows the relation between the annual variation of number of events and mean sunspot numbers. The solar maximum of cycle 23 falls in the year 2000 whereas solar maximum of cycle 24 falls in year 2012. From the figure it is clear that from year 2007 to 2009 there is a deep solar minimum as discussed by several investigators. Due to this deep solar minimum there is no DH CME in year 2009 and only 03 DH CMEs occur in 2007 and only one event in 2008. There is a good annual variation between the events.

DISCUSSION AND CONCLUSIONS

The release mechanism of magnetic energy in the corona is a core issue of the CME studies. The CMEs speeds are given by the numerical differentiation using interpolation of three lagrangian points. The uncertainties of the CME speed and acceleration come mainly from the uncertainty in height measurements (Song et al., 2013). The average speed of 471 DH CMEs is 1043 km/s and their median is 943 km/s. The DH CMEs linear speeds distribtution is in the range from 112 km/s up to 3387 km/s. The average width of 471 DH CMEs is 235° and their median is 360° for the solar cycle 23 and 24. In solar cycle 24 (January 2009–June 2013), we have considered 95 DH CMEs, out of which 59 are full halo CMEs (width = 360°), i.e. 62% of the DH CMEs are full halo CMEs, and 75% are halo CME (width > 180°). The CME of solar cycle 24 expand anomalously compared to those in cycle 23. This is supported by a larger fraction of halo CME (width > °) in solar cycle 24 CMEs. This anomalous expansion of CMEs can be attributed to the significant reduction in the observed total pressure (Magnetic + plasma) in the ambient medium into which the CMEs are ejected (Gopalswamy et al., 2014). The acceleration of the 471 DH CMEs during the study period varies from 434 m/s^2 to −179 m/s^2. Most of the events are found to be decelerated (i.e. 62%) and 34% move with positive acceleration. There is a good annual variation between the average widths and average speed, both the curve almost move parallel (c.f. fig. 5), there is break in the year 2009, which distinguish between solar cycle 23 and 24. It also shows a good yearly distribution of number of events with the mean sunspot numbers (c.f. Fig. 6). In the end we conclude that these DH CMEs are relatively faster and wider than the normal CMEs, (Gopalswamy et al., 2005 and Sharma et al., 2008), which is relevant to space weather.

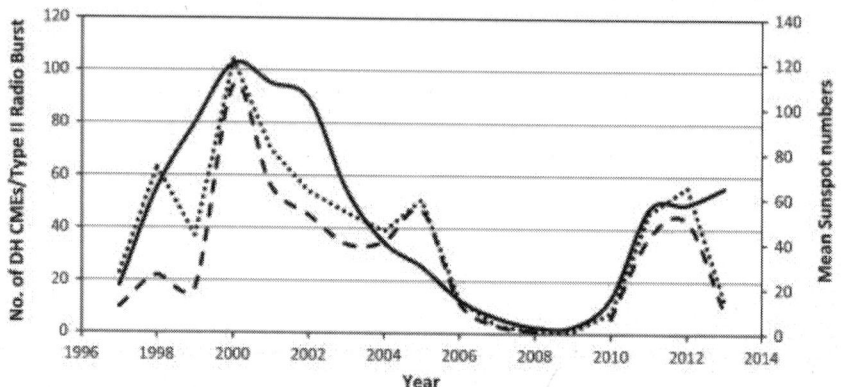

Figure 6: Annual variation of number of events and mean sunspot numbers. Solid line represents mean sunspot number; dashed line represents DH CMEs whereas dotted line represents type II events.

ACKNOWLEDGMENTS

Authors are also thankful to IUCAA, Pune for providing financial assistance, help and encouragement during the course of this work. The CME catalog is generated and maintained at the CDAW Data Center by NASA and The Catholic University of America in cooperation with the Naval Research Laboratory. SOHO is a project of international cooperation between ESA and NASA. The authors would like to thank for the excellent LASCO-CME catalogue, which includes supportive data the back-bone of our paper. Authors are also thankful to Wind/WAVES Spacecraft organization for providing the data.

REFERENCES

Acuna, M.H., Ogilvie, K.W., Baker, D.N., Curtis, S.A., Fairfield, D.H., Mish, W.H., 1995. The global geospace science program and Its Investigations. Space Sci. Rev. 71, 5.

1. Bougeret, J.L., Kaiser, M.L., Kellogg, P.J., Manning, R., Goetz, K., Monson, S.J., Monge, N., Friel, L., Meetre, C.A., Perche, C., Sitruk, L., Hoang, S., 1995. Waves: the radio and plasma wave investigation on the wind spacecraft. Space Sci. Rev. 71, 231.
2. Bruckner, G.E., Howard, R.A., Koomen, M.J., et al, 1995. The large angle spectroscopic coronagraph (LASCO). Sol. Phys. 162, 357.
3. Cane, H.V., Sheeley Jr., N.R., Howard, R.A., 1987. Energetic interplanetary shocks, radio emission, and coronal mass ejections. JGR 92, 9869.
4. Cho, K.S., Bong, S.C., Moon, Y.J., Shanmugaraju, A., Kwon, R.Y., Park, Y.D., 2011. Relationship between multiple type II solar radio bursts and CME observed by STEREO/SECCHI. Astron. Astrophys. 530, A16.
5. Cliver, E.W., Webb, D.F., Howard, R.A., 1999. On the origin of solar metric type II bursts. Sol. Phys. 187, 89.
6. Gopalswamy, N., Yashiro, S., Kaiser, M.L., Howard, R.A., Bougeret, J.L., 2001. Characteristics of coronal mass ejections associated with long-wavelength type II radio bursts. JGR 106, 29219.
7. Gopalswamy, N., Aguilar-Rodriquez, E., Yashiro, S., Nunes, S., Kaiser, M.L., Howard, R.A., 2005. Type II radio bursts and energetic solar eruptions. JGR 110, A12S07.
8. Gopalswamy, N., Thompson, W.T., Davile, J.M., Kaiser, M.L., Yashiro, S., Makela, P., Michalek, G., Bougeret, J.L., Howard, R.A., 2009. Relation between type II bursts and CMEs inferred from STEREO observations. Sol. Phys. 259, 227.
9. Gopalswamy, N., Akiyama, S., Yashiro, S., Xie, H., Makela, P., Michalek, G., 2014. Anomalous expansion of coronal mass ejections during solar cycle 24 and its space weather implications. Geophys. Res. Lett. 41, 2673.
10. Khan, J.I., Aurass, H., 2002. X-ray observations of a large-scale solar coronal shock wave. A&A 383, 1018.
11. Magdalenic, J., Vrsnak, B., Pohjolainen, S., Temmer, M., Aurass, H., Lehtinen, N.J., 2008. A flare-generated shock during a coronal mass ejection on 24 December 1996. Sol. Phys. 253, 305.
12. Nelson, G.J., Melrose, D.B., 1985. Type II Bursts; IN: Solar Radiophysics: Studies of Emission from the Sun at Metre Wavelengths (A87-13851 03-92). Cambridge University Press, Cambridge and New York, pp. 333.
13. Sharma, J., Mittal, N., Tomer, V., Narain, U., 2008. On properties of radio-rich coronal mass ejections. Astrophys. Space Sci. 317, 261.

14. Song, H.Q., Chen, Y., Ye, D.D., Han, G.Q., Du, G.H., Li, G., Zhang, J., Hu, Q., 2013. A study of fast flareless coronal mass ejections. Astrophys J 773, 129.
15. Subramanian, K.R., Ebenezer, E., 2006. A statistical study of the characteristics of type II doublet radio bursts. A&A 451, 683.
16. Uchida, Y., 1960. On the exciters of type II and type III solar radio bursts. Publ. Astron. Soc. Jpn. 12, 376. Vrsnak, B., Ruzdjak, V., Zlobec, P., Aurass, H., 1995. Ignition of MHD shocks associated with solar flares. Sol. Phys. 158, 331.
17. Wild, J.P., McReady, L.L., 1950. Observations of the spectrum of high-intensity solar radiation at metre wavelengths. I. The apparatus and spectral types of solar burst observed. Aust. J. Sci. Res. A3, 387.
18. Yashiro, S., Gopalswamy, N., Makela, P., Akiyama, S., Uddin, W., Srivastava, A.K., Joshi, N.C., Chandra, R., Manoharan, P.K., Mahalakshmi, K., Dwivedi, V.C., Jain, R., Awasthi, A.K., Nitta, N.V., Aschwanden, M.J., Choudhary, D.P., 2014. Homologous flare-CME events and their metric type II radio burst association. Adv. Space Res. 54 (9), 1941.

CITATION

Joginder Sharma, Nishant Mittal, Udit Narain, On some statistical characteristics of radio-rich CMEs in the solar cycles 23 and 24, NRIAG Journal of Astronomy and Geophysics, Available online 3 April 2015, ISSN 2090-9977, http://dx.doi.org/10.1016/j.nrjag.2015.03.001.

Chapter 9

Radiation Hydrodynamics Using Characteristics on Adaptive Decomposed Domains for Massively Parallel Star Formation Simulations

Lars Buntemeyer [a], Robi Banerjee[a] , Thomas Peters [b,c] , Mikhail Klassen[d], Ralph E. Pudritz [e]

[a] Hamburger Sternwarte, Universität Hamburg, Gojenbergsweg 112, Hamburg 21029, Germany
[b] Institut für Computergestützte Wissenschaften, Universität Zürich, Winterthurerstrasse 190, Zürich CH-8057, Switzerland
[c] Max-Planck-Institut für Astrophysik, Karl-Schwarzschild-Str. 1, Garching D-85748, Germany
[d] Department of Physics and Astronomy, McMaster University, 1280 Main Street W, Hamilton, ON L8S 4M1, Canada
[e] Origins Institute, McMaster University, 1280 Main Street W, Hamilton ON L8S 4M1, Canada

ABSTRACT

We present an algorithm for solving the radiative transfer problem on massively parallel computers using adaptive mesh refinement and domain decomposition. The solver is based on the method of characteristics which requires an adaptive raytracer that integrates the equation of radiative transfer. The radiation field is split into local and global components which are handled separately to overcome the non-locality problem. The solver is implemented in the framework of the magneto-hydrodynamics code FLASH and is coupled by an operator splitting step. The goal is the study of radiation in the context of star formation simulations with a focus on early disc formation and evolution. This requires a proper treatment of radiation physics that covers both the optically thin as well as the optically thick regimes and the transition region in particular. We successfully show the accuracy and feasibility of our method in a series of standard radiative transfer problems and two 3D collapse simulations resembling the early stages of protostar and disc formation.

INTRODUCTION

Radiative feedback plays a crucial role in the process of star and disc formation, the evolution of circumstellar discs and the thermodynamics of the interstellar medium (ISM). Massive stars emit a large number of energetic UV photons and strongly determine the structure of giant molecular clouds (GMCs) by creating large bubbles of ionized gas (HII regions) (e.g. Peters, Banerjee, Klessen, Mac Low, Galván-Madrid, Keto, 2010, Walch, Whitworth, Bisbas, Wünsch, Hubber, 2012 and Dale, Ercolano, Bonnell, 2013). On smaller scales, low mass and intermediate mass stars also significantly influence their surroundings by radiative heating. By increasing the fragmentation scale, radiative heating can completely inhibit further fragmentation in a radius of several AU and prevent, e.g., the formation of a binary system (Price and Bate, 2010). Offner et al. (2009) investigate the initial mass function (IMF) and the star formation rate (SFR) by comparing 3D hydrodynamical simulations of low mass star formation with and without the effects of radiative transfer. They find that the thermal support of a protostar's accretion luminosity suppresses further fragmentation in the cloud core as well as in the protostellar disc. The SFR in their simulations is about half the value of the simulations without radiative transfer and the mass distribution of protostars of very low mass ($M_* < 0.1 \, M_\odot$) is significantly reduced. Bate (2009) finds similar effects.

Regarding the formation and evolution of circumstellar discs, radiative feedback is indispensable to understand their fragmentation behavior, thermodynamics, and morphology (Chiang and Goldreich, 1997) and to model the infrared excess observed in their spectral energy distributions (SEDs) (e.g. Dullemond and Monnier, 2010). The initial formation of massive discs during the Class 0 phase has been investigated using hydrodynamical and magnetohydrodynamical (MHD) simulations (e.g. Yorke, Bodenheimer, Laughlin, 1993, Mellon, Li, 2008, Machida, Inutsuka, Matsumoto, 2010,Peters, Banerjee, Klessen, Mac Low, Galván-Madrid, Keto, 2010 and Seifried, Banerjee, Klessen, Duffin, Pudritz, 2011), and Seifried et al. (2013) emphasize the importance of turbulence to explain the formation of Keplerian discs even if strong magnetic fields are present.

Despite a large number of studies, the actual transition from the early self-gravitating protostellar disc (Class 0) to the Keplerian protostellar disc is still poorly understood. Recent observations (e.g. Tobin et al., 2012) indicate that Keplerian discs might form very early during the protostellar evolution and the analytic study by Forgan and Rice (2013) emphasizes the effects of radiative processes. However, the effects of radiative transfer have usually been neglected in MHD simulations so far or were substantially approximated (e.g. Yorke, Bodenheimer, Laughlin, 1993, Mellon, Li, 2008, Machida, Inutsuka, Matsumoto, 2010 and Seifried, Banerjee, Klessen, Duffin, Pudritz, 2011). The self-consistent modeling of the formation and early evolution of protostars and protostellar discs therefore creates the need for numerical methods to make 3D radiation MHD simulations feasible.

In this context, radiative transfer is a rather costly computation compared to solving Euler's equations. The reason for this is that the timescale of radiative transfer is usually much shorter than those of hydrodynamics and MHD because of the large speed of light compared to the sound speed of the gas in, e.g., a molecular cloud or the characteristic Alfvén wave speeds of the magnetic field. The short timescale on which radiation emerges throughout the complete computational domain makes radiative transfer a highly non-local problem compared to MHD which is determined completely by local thermodynamic properties of the gas. In this sense, hydrodynamics and radiative transfer are two very different numerical tasks and very challenging to solve consistently.

Modern Eulerian MHD codes like FLASH (Fryxell et al., 2000) mostly solve the Euler equations on a grid with adaptive mesh refinement (AMR) to resolve fluid features on a wide range of length scales. These codes are parallelized by subdividing the computational domain into several subdomains each containing a fixed number of cells. Since the Euler equations describe local fluxes of mass, momentum and energy, all subdomains can be handled in parallel during a hydrodynamical time step. Between the time steps, boundary values of the subdomains are exchanged using the Message Passing Interface (MPI) for communication. In contrast, characteristics based radiative transfer codes are usually designed very differently. Instead of domain decomposition, these codes are parallelized exploiting the formal independence of the radiative transfer equation (RTE) on the solid angle. Resolving the anisotropy of the radiation field accurately requires a large set of characteristics each covering a discrete opening angle of the 4π unit sphere. If all radiative quantities are assumed to be fixed during one solution step, characteristics of different directions can be computed independently of each other which makes it ideal for parallelization. However, the spatial information of the computational domain with all radiative quantities has to be available for each processor computing a certain number of characteristics on the solid angle grid. This can be a severe drawback in terms of memory requirement if high spatial resolution is required or a large number of frequencies or both (e.g., synthetic stellar spectra). Solving both Euler's equations and the RTE consistently requires careful approximations to the radiative transfer problem to make the coupling of an MHD code with a radiative transfer code feasible. van Noort et al. (2002) present a radiation solver that is coupled to a hydrodynamical code using AMR and domain decomposition in 2D. The radiation solver uses short characteristics (SC) for integrating the RTE while boundary values are communicated between Lambda iteration steps. The focus of this approach lies on modeling the dynamics of scattering dominated stellar atmospheres. The SC approach allows for a fast converging Gauss–Seidel iteration scheme (e.g., Trujillo Bueno and Fabiani Bendicho, 1995), while non-local contributions have to be communicated by a successive exchange of boundary values between subdomains. This approach was also extended for 3D simulations (e.g., Hayek, Asplund, Carlsson, Trampedach, Collet, Gudiksen, Hansteen, Leenaarts, 2010 and Davis, Stone, Jiang, 2012). However, while the Gauss–Seidel short characteristics approach is well suited for highly scattering dominated regimes, it introduces a lot of numerical diffusion because a large number of upwind interpolations is necessary. Razoumov and Cardall (2005) implement a method that is as computationally cheap as the SC method but less diffusive. They create rays on each refinement level separately while their approach is fully threaded but not MPI parallelized. Recently, Tanaka et al. (2014) parallelized this approach using the *multiple wavefront method* byNakamoto et al. (2001) based on a carefully chosen calculation sequence on a spatially decomposed domain. This method requires successive communication of boundary values. A similar approach is used with long characteristics (LC) in 3D by Heinemann et al. (2006) without AMR.

Another approach for including radiative transfer in hydrodynamical simulations is based on the moment equations (the angular integrated RTE) of the zeroth, first and second moment of the specific intensity. A moment-based scheme does not necessarily require to integrate along large sets of characteristics, however, the anisotropy of the radiation field has to approximated reasonably in order to close the set of moment equations for the mean intensity, radiative flux and pressure. A possible closure relation is the M1-closure used, e.g., in the HERACLES code (González et al., 2007). The closure relation can also be explicitly calculated using, e.g., a characteristics based approach which is known as the Variable Eddington Tensor (VET) method (e.g. Jiang et al., 2012). A common approach for star formation simulations is the diffusion approximation of the angular moment equations which assumes the radiation field to be completely isotropic. In regions of high opacities χ, the diffusion approximation is an expansion of the specific intensity in which all terms $\propto 1/\chi$ are neglected in the RTE. This leads to Eddington's approximation in which the isotropic radiation pressure is proportional to the radiation energy density. The moment equations of the radiative intensity themselves then form a set of hyperbolic equations, like Euler's equations. However, since those two hyperbolic systems would still have to be handled on their individual timescales, one can even make one further step and neglect the time dependence of the radiation flux by assuming it to be proportional to the gradient of the radiation energy (Fick's law). The moment equations can then be combined into a single diffusion equation for the energy of the radiation field. Because the flux in the diffusion approximation lost its finite propagation speed, one has to introduce a flux-limiter to avoid unphysical propagation speeds depending on the actual opacity. This *flux-limited diffusion approximation* (FLD) (Levermore and Pomraning, 1981) has been successfully used in radiation hydrodynamical star formation simulations coupled within Eulerian grid codes (e.g.Stone, Mihalas, Norman, 1992, Krumholz, Klein, McKee, 2007, Commerçon, Teyssier, Audit, Hennebelle, Chabrier, 2011, Flock, Fromang, González, Commerçon, 2013,Zhang, Tan, McKee, 2013 and Bryan, Norman, O'Shea, Abel, Wise, Turk, Reynolds, Collins, Wang, Skillman, Smith, Harkness, Bordner, Kim, Kuhlen, Xu, Goldbaum, Hummels, Kritsuk, Tasker, Skory, Simpson, Hahn, Oishi, So, Zhao, Cen, Li, The Enzo Collaboration, 2014) as well as smoothed particle hydrodynamics (SPH) codes (e.g. Bate et al., 2013). However, the diffusion approximation is only valid in optically thick regions where the radiation field becomes isotropic. Kuiper et al. (2010) have shown the significant drawbacks of exclusively using FLD in the transition regions from optically thick to optically thin regimes where the radiation field becomes highly anisotropic. Recent efforts have been made to combine raytracing methods with FLD solvers (Flock, Fromang, González, Commerçon, 2013, Klassen, Kuiper, Pudritz, Peters, Banerjee, Buntemeyer, 2014 and Kuiper, Klahr, Dullemond, Kley, Henning, 2010) to handle, at least, primary stellar or protostellar radiation separately from the FLD approximation and to avoid the stellar flux from diffusing into shadow regions.

Finally, Monte Carlo (MC) methods have become increasingly popular during the last decade, especially in post-processing MHD simulations. The MC method is a statistical approach and treats individual photons or photon packages by following its propagation path and computing absorption, emission and scattering probabilities. Several advances have been introduced, e.g., photon peel-off (Lucy, 1999), immediate reemission (Bjorkman and Wood, 2001) and diffusion approximations (Min et al., 2009) which make the MC method a powerful tool to calculate synthetic spectra, SEDs or polarization maps from the outcome of MHD simulations. The angular and frequency resolutions are, in principle, unlimited since the direction of propagation of a photon package and its frequency are chosen randomly from a continuous probability function. In that sense, the MC method always gives a quite reasonable result even in the limit of a small number of photon packages while a low resolution shows mainly up as statistical noise in the solution. But the statistical approach also has a severe drawback since we do not know in advance the exact path a photon package will travel, and how and when it is emitted or absorbed. Therefore, it is extremely difficult to implement on a decomposed domain. MC methods are extremely successful in post-processing the outcome of MHD simulations but are rarely used in combination with hydrodynamical simulations. Those approaches which does include MC methods (e.g. Acreman et al., 2010) are fairly restricted in their spatial resolution of the AMR grid, since each processor has to get a copy of the complete computational domain to be able to follow the path of an arbitrary photon package. For our approach, we therefore choose a discrete ordinate method using characteristics to integrate the RTE which requires a raytracer that works on an AMR grid with domain decomposition.

The paper is organized as follows: In Section 2, we give a brief introduction into the theory of radiative transfer as far as it concerns our method and we describe in detail the method of hybrid characteristics. We also describe the coupling between our radiation solver and the FLASH code. In Section 3, we show results from test calculations we performed to investigate the accuracy of our radiation code as well as the coupling to the FLASH code. In Section 4, we present results from 3D radiation hydrodynamical collapse simulations and the parallel scaling performance of our code is described in Section 5. InSection 6, we discuss our results and put it into context with other state-of-the-art radiation transfer methods.

THEORY AND NUMERICS

In this section, we describe the theory of radiative transfer (RT) that forms the basis of our solution method as well as the numerics. We describe the hydrodynamics only as it becomes important in the coupling with the radiation solver. For a more detailed description of the FLASH code and its capabilities we refer to Fryxell et al. (2000).

The Equation of Radiative Transfer

The theory of radiative transfer in this section is based on the work by Mihalas and Weibel Mihalas (1984) in the limit of geometrical optics. The energy of the radiation field is described by a scalar field of specific intensities $I(\mathbf{x}, \mathbf{n}(\vartheta, \varphi), v)$, where $\mathbf{x}$ is the position in space, ϑ and φ define the direction of propagation $\mathbf{n}$, and v is the frequency. The radiative transfer equation (RTE) describes the change of the specific intensity during its propagation in a medium which is determined by an energy balance between emission and absorption processes. It reads

$$\frac{1}{c}\frac{\partial I_v}{\partial t} + \mathbf{n} \cdot \nabla I_v = \eta_v - \chi_v I_v,$$

(1)

where η_v denotes the emissivity (energy volume density per unit time and solig angle), χ_v is the extinction coefficient (1/length) and c is the speed of light. The specific intensity denotes the radiative energy flux *per solid angle* $d\Omega = \sin\vartheta\, d\vartheta\, d\varphi$, thus in vacuum, it is constant along a line of sight.

Interaction processes between radiation and matter determine the extinction coefficient χ_v. However, for this work, we solve the time independent RTE since we assume the radiation field to emerge instantaneously throughout the entire computational domain during a hydrodynamical time step. Furthermore, we use the definition of the *source function* $S_v = \eta_v/\chi_v$ and rewrite the RTE in terms of the optical depth without explicit frequency dependence:

$$\frac{dI(\mathbf{n})}{d\tau(\mathbf{n})} = S - I(\mathbf{n}).$$

(2)

This form of the RTE describes the propgation of the specific intensity along a specific line element ds in the direction $\mathbf{n}$ and the optical depth element $d\tau = \chi\, ds$ respectively. This requires a proper paramerization depending on the coordinate system and, hence, the definition of $\mathbf{n} \cdot \nabla$. Note that the RTE in the form of Eq. (2)becomes a 1D ordinary differential equation. However, for the numerical solution in 3D, the optical depth element $d\tau$ is discretized and parameterized in Cartesian coordinates and the solution is obtained by integrating the RTE on a solid angle grid. The source function S is a more general form of Kirchoff's law. It describes the ratio of emission and extinction of radiative energy and allows arbitrary contributions from thermal emission as well as scattering contributions. In fact, the complexity of the model and the solution of the RTE depend strongly on the source function we choose to accurately describe the current radiation transfer problem (e.g. local thermodynamic equilibrium (LTE), non-LTE (NLTE), grey or non-grey, anisotropy, dust continuum radiation, line transfer, etc.). Describing the complete radiation field would require 6 dimensions, which makes it an extremely challenging task to compute and store a 3D solution. In order to handle the radiation field numerically, the intensity is only computed on the fly and accumulated in form of the solid angle averaged mean intensity

$$J = \frac{1}{4\pi} \oint_{4\pi} I \, d\Omega.$$

(3)

The mean intensity is the zeroth moment of the specific intensity and closely related to the radiative energy density $E_r = \frac{4\pi J}{c}$. Depending on the model setup, the source function usually depends on the radiation field itself and the mean intensity becomes a part of the source function, which makes the RTE an integro-differential equation. In order to find a self-consistent solution, one has to invoke an iteration scheme of some form. Formally however, the RTE from Eq. (2) can be solved by the *formal solution*:

$$I(\tau_2) = I(\tau_1) \, e^{(\tau_2 - \tau_1)} + \int_{\tau_1}^{\tau_2} S(\tau) \, e^{(\tau_2 - \tau)} d\tau.$$

(4)

The formal solution describes the intensity propagation along a line element with the optical depth $\Delta\tau = \tau_2 - \tau_1$. It contains the incoming intensity $I(\tau_1)$, which is partially extinct and additional energy from emission processes. The RTE and the formal solution are linear in the intensity and allow us to split the radiation field in as many components as the solution method requires. This is a crucial part in our approach of solving the RTE on a decomposed computational domain such as the adaptive grid embedded in the FLASH code.

Numerical Radiative Transfer

Integrating the RTE along a set of rays of different directions **n** using Eq. (4) is based on the *method of characteristics*. It was first introduced into the radiation transfer community by Olson et al. (1986). The RTE is integrated for each cell in the computational domain and each direction by computing a stepwise formal solution along a ray, or *long characteristic* (LC), according to the discretized formal solution

$$I_i = I_{i-1} \exp(-\Delta\tau_{i-1}) + \Delta I_i, \tag{4}$$

where $\Delta\tau_i$ is the finite optical depth element given by a piecewise linear interpolation

$$\Delta\tau_i = \frac{1}{2}(\chi_{i-1} + \chi_i)\Delta s. \tag{5}$$

χ_i is the opacity at the discretization point s_i on the characteristic. ΔI_i is the discretized counterpart to the integral in the formal solution (Eq. (4)) and is solved by either a linear or parabolic interpolation according to

$$\Delta I_i = \alpha_i S_{i-1} + \beta_i S_i + \gamma_i S_{i+1}. \tag{6}$$

The coefficients α_i, β_i and γ_i depend on the optical depths between s_{i-1}, s_i and s_{i+1}. They are given in Olson et al. (1986). Fig. 1 shows the geometrical situation of a characteristic passing through a homogenous grid at an arbitrary direction $n_j(\vartheta, \varphi)$. Since the opacity and source function are stored in the cell centers of the finite volume FLASH grid (dashed lines), they are assumed to be constant inside the cell. However, since we use a parabolic interpolation [1] of the source function integral, we introduce a point-based RT grid which is based on the cell centers of the FLASH grid. These cell centers define the vertices from which we interpolate the values at the intersection points of the ray bilinearly. Consequently, the point-based RT grid is staggered by half a grid cell since the ray does not intersect with the grid faces of the finite volume grid but with the faces of the point based grid defined by the FLASH cell centers. The characteristic is traced on the RT-grid using the fast voxel traversal algorithm introduced by Amanatides and Woo (1987). The opacity χ_i and the source function S_i at the intersection points of the characteristic with the RT-grid are interpolated bilinearly from the adjacent vertices.

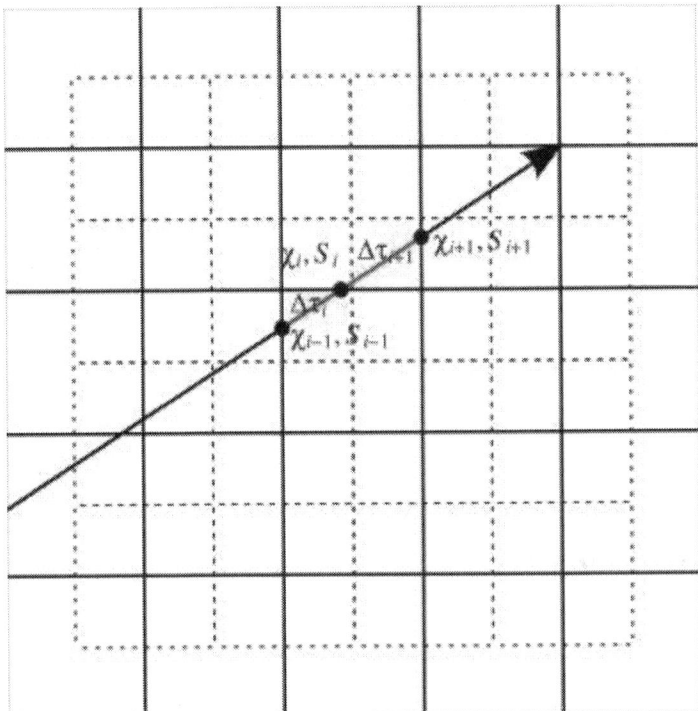

Figure 1. The staggered RT-grid (solid lines) defined by the cell centers of the underlying finite volume hydro-grid (dashed lines); a long characteristic at an arbitrary direction is shown, which integrates the RTE for the hydro cell-center at the upper right corner of the domain.

While the RTE is integrated along each direction **n**, the mean intensity is computed by accumulating all intensities:

$$ J = \frac{1}{4\pi} \sum_{\mathbf{n}} I(\mathbf{n}) \Delta\Omega, $$

(7)

which requires a discretization of the solid angle Ω. If no information about the anisotropy of the radiation field is available, one should choose a homogeneous discretization which is a non-trivial problem if one considers spherical coordinates on the unit sphere. For this purpose, we use the HEALPix (Hierarchical Equal Area isoLatitude Pixelization) scheme introduced by Górski et al. (2005). HEALPix ensures an optimal discretization of the unit sphere into a number of finite solid angles $\Delta\Omega$. The discretization is based on 12 base pixels which are subdivided depending on the required resolution level. Consequently, typical numbers of directions for the integration of the specific intensity are $N_{pix} = 12, 48, 192$ or 768 (Appendix B)Characteristics based radiative transfer is the attempt to approximate the radiative interaction of each cell with each other cell in the computational domain. Although the method of long characteristics is very accurate in doing this, it is rather inefficient as it requires to shoot a large number of rays for each cell to sample the radiation field accurately in 3D. An alternative is to use a short characteristics (SC) approach, in which only neighboing cells are used to interpolate incoming intensities from different directions. This requires to sweep the cells in an ordered fashion to make sure that all intensities which are required for interpolation are available. The SC approach introduces numerical diffusion because of the large number of interpolations involved but reduces the cost of the RT calculations by a factor of n_c (the total number of cells involved). Either way, the method of characteristics must invoke a raytracer, which samples radiative interactions between arbitrary regions in the computational domain.

Raytracing on the Decomposed AMR Grid

The parallel design of the FLASH framework, in which our solver is currently implemented, forbids to trace rays over the entire domain as it is necessary for the method of characteristics. FLASH invokes PARAMESH (Olson et al., 1999), and lately also the CHOMBO library,[2] for implementing an adaptive mesh refinement (AMR) grid. Paramesh uses a *block* structured AMR mesh, in which the fundamental data structure is a block containing cells which are logically indexed by a coordinate triple (*i,j,k*). The entire computational domain consists of a number of blocks of different physical sizes ordered hierarchically in an octree data structure. Blocks at the bottom of the tree structure, called *leaf blocks*, contain valid data and they cover the entire physical size of the computational domain. FLASH allows for massively parallel computation by invoking the Message Passing Interface (MPI) for the communication of ghost cell information between the blocks. Optimal load balancing is guaranteed by splitting the AMR tree equally between all available MPI tasks to ensure that each task receives more or less the same number of leaf blocks. E.g., the AMR tree of a star formation simulation typically requires more than 10 levels of spatial resolution with up to several 10^5 blocks each containing 8^3 cells, which is only made possible (in terms of cpu-time and memory requirements) by the parallelization described above.

The method of characteristics stays in direct contrast to the spatial parallelization of the AMR grid (Fig. 2). However, in order to account for non-local coupling of the radiation field, we adapt a raytracing technique originally developed by Rijkhorst et al. (2006) and improved by Peters et al. (2010), which uses a combination of local long characteristics and a global "short-characteristics-like" interpolation of outgoing intensities from the decomposed domains of the AMR grid. The basic idea is to split the radiation field in two components:

- A local component uses long characteristics to compute only radiative contributions to both the cells inside a block as well as contributions that leave the block (*face values*). The computation is done in parallel and in accordance with the design of the block structured AMR grid.
- A global component which is computed by communicating and accumulating face values (see Fig. 3). This step invokes raytracing over the block structure of the AMR tree and a linear interpolation of face values very similar to the SC method (but on the level of subdomains). After the communication of face values and the tree hierarchy, this step is also done in parallel.

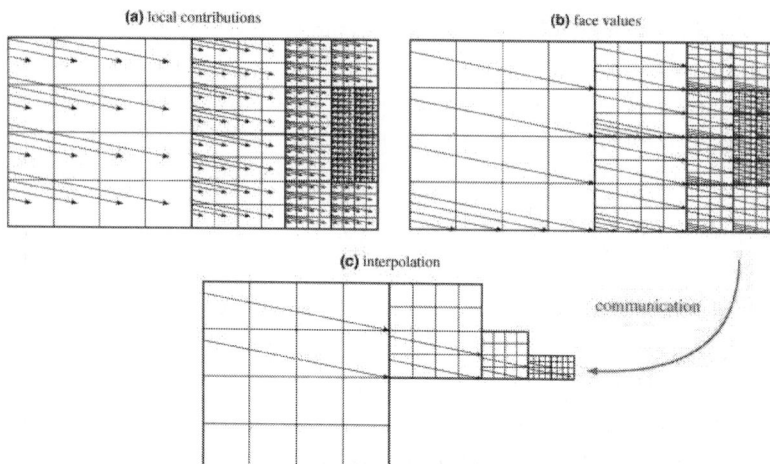

Figure 3. The basic steps of the hybrid characteristics method for parallel rays (compare to Rijkhorst et al. (2006)) in a 2D AMR domain that is refined from left to right. Bold lines show the boundaries of the patches at different AMR levels (in FLASH, these patches are called *blocks* and are distribted equally over the available number of MPI tasks). In this example, each block contains 4x4 cells (indicated by thin lines). (a) Local contributions as calculated with long characteristics. (b) The outgoing face values which are communicated. In fact, we communicate all face values even though they might be part of the same subdomain of a certain MPI process since we need them in the following interpolation step. (c) Example for the linear interpolation of face values for a particular target cell after the communication step. The linear interpolation requires to weight the face values from two rays. The weights depend on where a certain ray segment starts at the boundary of a block (which is a 4x4 cell patch at a certain refinement level) or subdomain respectively.

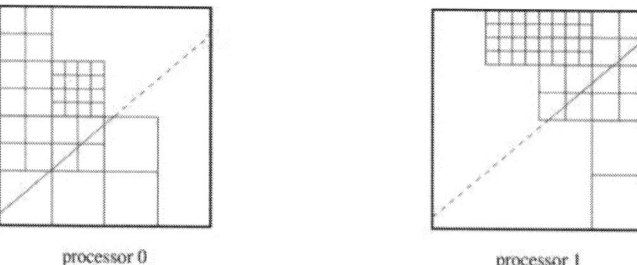

processor 0   processor 1

Figure 2. Example for a 2D AMR grid distributed over two processors without shared memory. The bold rectangle shows the boundary of the whole computational domain. Thin lines show the leaf blocks at different refinement levels that make up the whole subdomain a processor is assigned to. Raytracing through the domain is obviously restricted to the subdomain each processor is assigned to.

This approach, called *hybrid characteristics*, only needs to communicate the face values of the blocks and information about the AMR tree hierarchy but no 3D data. By this, the amount of communicated data is reduced significantly. Originally, this method was developed by Rijkhorst et al. (2006) to compute column densities only with respect to point sources for UV ionization. The original method requires to communicate the whole AMR tree structure at the highest level of spatial resolution during the raytracing step on the AMR block structure. This stands in contrast to the parallel design of the FLASH code and restricts the available range of refinement levels of the AMR tree substantially because of the large memory overhead. Peters et al. (2010) add some major improvements to the algorithm by introducing a walk through the AMR tree, which only requires the communication of basic AMR information and conserves the idea of shared memory parallelization.

However, the method was, originally, restricted to compute column densities along rays which originate at a certain point in the grid and used to represent, e.g., a stellar source. For this work, we removed this restriction and implemented a radiative transfer framework which is able to compute the radiation field independently from any point source by solving the RTE for large sets of characteristics along parallel rays. By combining our improvements with the original method, the solver cannot only account for the primary emission by point sources (as in Rijkhorst et al. (2006) and Peters et al. (2010)) but also for the reemitted, diffuse component of the radiation field. Fig. 4 shows a 2D example of a simple test setup with an irradiated central density clump using AMR. From the figure we can see the ability of the method to create sharp shadows and to transport incoming radiation over the entire domain.

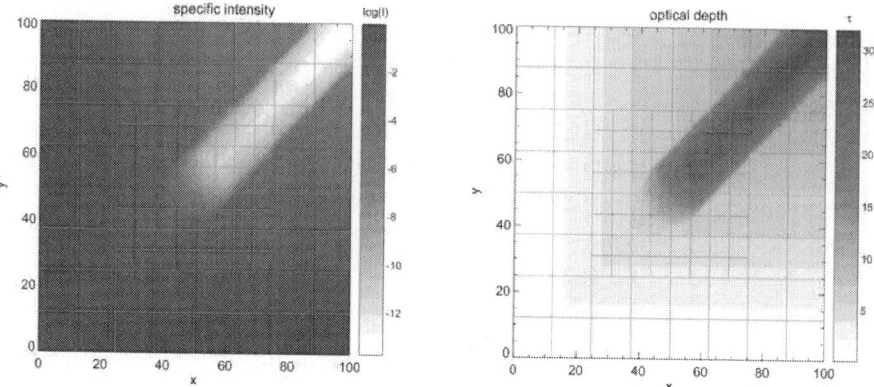

Figure 4. The specific intensity and the optical depth computed diagonally in the *xy*-plane with a central density clump. The source function is set to unity at the left and bottom outermost boundaries and zero everywhere else. The opacity of the central clump is one order of magnitude higher than the ambient opacity. The grid indicates the block structure of the AMR grid, units are arbitrary.

Coupling to the FLASH Code

Since our method is implemented in the FLASH framework, it is straightforward to couple the radiative transfer module to the hydrodynamical and MHD modules of the FLASH code. The coupling is done by accounting for radiative emission and absorption processes, which are determined by the thermal emission opacity $\chi_e = \kappa_e \rho$ and the thermal absorption opacity $\chi_a = \kappa_a \rho$. The opacities are calculated from mass specific cross sections κ_e and κ_a. Note that the total extinction coefficient χ, which is used for the solution of the RTE, may include an additional scattering opacity χ_s and therefore $\chi = \chi_a + \chi_s$. The coupling of both the radiation and the MHD solver is achieved by computing a source term according to Mihalas and Weibel Mihalas (1984) which describes the total net gain or loss of energy due to radiative heating and cooling. It reads

$$Q_{rad} = 4\pi\chi(J-S) = 4\pi\chi_a(J-B).\qquad(8)$$

This source term is computed from the time-independent solution of the radiation field as described in the previous sections and it is coupled to the MHD integrator in an operator splitting step. Hence, the set of compressible MHD equations in dimensionless form including gravitation and radiative energy exchange are those of

continuity

$$\frac{\partial \rho}{\partial t} + \nabla \cdot (\rho \mathbf{v}) = 0,$$

(9)

momentum conservation

$$\frac{\partial (\rho \mathbf{v})}{\partial t} + \nabla \cdot (\rho \mathbf{v} \otimes \mathbf{v} + p_* 1 - \mathbf{B} \otimes \mathbf{B}) = \rho \mathbf{g}$$

(10)

energy conservation

$$\frac{\partial E}{\partial t} + \nabla \cdot (\mathbf{v}(E + p_*) - \mathbf{B}(\mathbf{v} \cdot \mathbf{B})) = \rho \mathbf{v} \cdot \mathbf{g} + Q_{\text{rad}},$$

(11)

and the induction equation

$$\frac{\partial \mathbf{B}}{\partial t} - \nabla \times (\mathbf{v} \times \mathbf{B}) = 0,$$

(12)

with the gas velocity field $\mathbf{v}$, the magnetic field vector $\mathbf{B}$ and the gravitational acceleration $\mathbf{g}$. p_* is the total pressure and E is the total energy density of a fluid element containing magnetic contributions according to

$$p_* = p + \frac{B^2}{2},$$

(13)

$$E = \frac{1}{2} \rho u^2 + e_{\text{int}} \rho + \frac{B^2}{2},$$

(14)

with the gas density ρ, the thermal pressure p and the internal specific energy e_{int}. 1 denotes the unity matrix. The MHD equations are closed by an ideal gas equation of state (EOS) which relates the internal energy to the thermal gas pressure according to

$$p = (\gamma - 1)\rho e_{int}. \tag{15}$$

We assume $\gamma = 5/3$ which corresponds to a mono atomic (hydrogen) gas. The temperature is also related to the internal energy by:

$$e_{int} = (\gamma - 1)^{-1} \frac{k_b}{\mu m_p} T, \tag{16}$$

where k_b is the Boltzmann constant, m_p the proton mass and μ the mean molecular weight of the gas.

Note that we solve the equations of MHD and RT successively by an operator splitting step and not simultaneously. Furthermore, for the following test cases, the thermal pressure dominates the hydrodynamics and it is several orders of magnitude larger than the radiation pressure, which we therefore neglect in the momentum Eq. (10). However, since our method explicitly computes the angular dependency of the radiation field, it is straightforward to couple it into the MHD equations.

Choosing the Time Step

The current coupling is done by an update of the internal gas energy e_{int} and temperature T respectively. Since we solve the time independent RTE, there is no update of the radiative energy or the source function during the solution of Euler's equations. Instead this is done in the following time step when the gas quantities have been updated. The update of the internal energy is done explicitly by

$$\Delta e_{int} = \Delta t \, Q_{rad}. \tag{17}$$

Due to the explicit update, we have to make some restrictions on the time step. The radiation field does not have an explicit influence on the CFL time step since the energy update is done after the solution of the

MHD equations. Instead, we compute a cooling time step which is chosen if it is shorter than any other time step from a FLASH module. The cooling time step is chosen so that the energy contribution Δe_{int} does not exceed a fixed percentage of the internal energy. Otherwise, if the time step Δt is chosen too large, the total radiative energy could become negative (e.g., $\Delta e_r > e$). This leads to the following time step restriction:

$$\Delta t_c = \min \left(\frac{e_{int}}{|\Delta e_{int}|} \right)_i k_c \, \Delta t_{CFL},$$

(18)

where k_c determines how much change in the internal energy is allowed, Δt_{CFL} is the CFL time step, and i denotes the indices of all grid cells in the computational domain. Because of the explicit energy update, the cooling time step is usually shorter than the CFL time step. So far, there is no subcycling involved and the FLASH code chooses a global minimum time step from all physics modules involved (including, e.g., self-gravity). The cooling time step highly depends on the absorption coefficient χ since it determines the optical depth of the medium and how much radiation is absorbed and emitted during a single time step. Typically the choice of $0.2 > k_c > 0.01$ is convenient as it produces accurate results (Section 3.3) and time steps about one oder of magnitude lower than the CFL time step.

The Lambda Formalism

Computing the radiation field in the form of the mean intensity in Eq. (7) requires a formal solution of the RTE in the way described above. Usually, this task is described in a rather compact form by using the *Lambda operator*:

$$J = \Lambda[S].$$

(19)

Formally, the Lambda operator for *one cell* in the computational domain contains the radiative contributions from each other cell. The construction of the operator would require to explicitly calculate the radiative coupling between a cell and each other cell. But this is far too costly concerning computation time and memory requirements. Instead, we do not construct the operator but we approximate the Lambda step fromEq. (19) by using the formal solution from Eq. (4) to compute the radiation field J from the source function S in the way described above. The accuracy of this approximation in a 2D or 3D computation depends crucially on the angular resolution, since it determines whether we actually "hit" each other cell during the angular integration of the mean intensity or not. We avoid this problem partly by calculating the radiation from point sources (e.g. a stellar source) explicitly for each cell by combining our method with the original hybrid characteristics method by Rijkhorst et al. (2006) andPeters et al. (2010). However, the Lambda step from Eq. 19 requires that we know the source function in advance. If we take the temperature from FLASH's hydro solver, we can compute the source function simply as being $S = B(T)$ then solve for the radiation field, couple it back to the hydro solver and we are done. This approach assumes the gas to be in a state of thermodynamical equilibrium but this is, of course, not always the case. If the radiation field is decoupled from the gas temperature, we do not know the source function in advance. The solution then requires some kind of iterative procedure to account for the non-local coupling of the radiation field with the gas. In the theory of radiative transfer, this iterative method is called *Lambda iteration*, which requires iterating over Eq. (19) until a self-consistent solution for $J(S)$ is found. Strictly speaking, even in the LTE case with $S = B(T)$, we have to iterate to find a temperature that is consistent with the internal energy of the gas since this determines the thermal emission. However, the Lambda iteration may need several hundreds of iteration steps, which is too costly and ineffective to be employed in a hydrodynamical simulation. One way of resolving this problem, is to partly solve Eq. (19) analytically by splitting the Lambda operator. These approaches, called *Accelerated Lambda iteration*(ALI), have been investigated and used extensively in the stellar atmosphere community (e.g. Trujillo Bueno and Fabiani Bendicho, 1995). We have implemented the most simple form of ALI, the local lambda operator, to solve radiative transfer problems even in regions of high optical depths and strong decoupling where the classical Lambda iteration usually fails (Appendix A).

TESTS

In this section, we show test results from the implementation of our radiation solver. The tests include time independent (Sections 3.1 and 3.2) as well as dynamical tests (Section 3.3) in 1D and 3D. We also show results from the combined FLASH/RT code in a series of 1D radiative shock calculations in Section 3.4.

1D Atmosphere

In the first test, we compute the radiation field in a grey, isothermal, scattering dominated 1D atmosphere. This test is typically used to verify a radiation solver's iterative performance and accuracy in a non-local thermodynamic equilibrium (NLTE) situation on a wide range of optical depths. It is also particularly useful to ensure that the solver accurately reproduces the diffusion limit in an optically thick regime, e.g., in the lower parts of the atmosphere. This test also requires the ALI method since the classical Lambda iteration fails to reproduce the solution in the case of strong scattering contributions. The amount of scattered radiation is quantified by the ratio of the thermal absorption coefficient to the total extinction coefficient according to

$$\epsilon = \frac{\chi_a}{\chi_a + \chi_s}, \tag{20}$$

where we neglect the frequency dependence and ϵ is the *photon destruction probability*. The grey source function in the atmosphere contains a thermal part and a scattering contribution, and it reads

$$S = \frac{\eta}{\chi} = \frac{\eta_s}{\chi_a + \chi_s} + \frac{\eta_e}{\chi_a + \chi_s}, \tag{21}$$

$$= (1 - \epsilon)J + \epsilon B, \tag{22}$$

where we defined $J = \eta_s/\chi_s$, the thermal emission is $B = \eta_e/\chi_a$, and η_s and η_e denote the scattering and the thermal emissivity respectively. Since the atmosphere is isothermal, we assume that we know the temperature and normalize it so that B = 1. The crucial part in this test is to find the source function which has to be consistent with the mean intensity J which is

$$J = \frac{1}{2}(I_- + I_+),$$

(23)

where I_- and I_+ are the downward and upward (2 stream solution) integrated specific intensities respectively. Since we assume a uniform mass specific opacity κ and constant temperature T, the intensity is only a function of optical depth $d\tau = \chi dz$, thermal emission B and the photon destruction probability ϵ. The mean intensity is then given by the analytic solution

$$J = B\left(1 - \frac{\exp(-\sqrt{\epsilon}\tau)}{1 + \sqrt{\epsilon}}\right).$$

(24)

The density ρ of the model increases exponentially with distance from the upper boundary and we assume that $\chi \propto \rho$ but with ϵ being constant. There is no incoming radiation at the upper boundary of the atmosphere at $\tau = 0$ while at the lower boundary the incoming radiation is $I = B$. The resulting model atmosphere provides an exponentially varying optical depth τ which resolves the transition region from the optically thick inner LTE-regions to the optically thin NLTE-regions at the outer boundary. We test the solver for a wide range of photon destruction probabilities from $\epsilon = 10^{-1}$ to 10^{-8}. The domain consists of 8 subdomains each containing 8 cells which results in a total spatial resolution of 64 cells. Fig. 5 shows the results. In the outer optically thin parts of the atmosphere, the scattering contribution in the source function becomes dominant since radiation leaves the atmosphere. The numerical solution is in excellent agreement with the analytic solution.

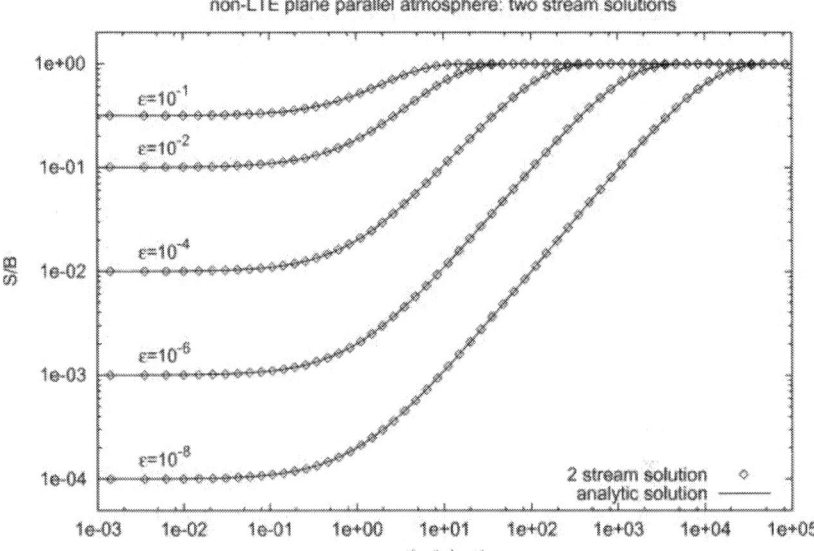

Figure 5. Scattering dominated 1D atmosphere problem. The solutions from the radiation solver (symbols) are compared to the analytic solutions (lines) for five different photon destruction probabilities.

Hydrostatic Protostellar Disc

Cosmic dust is one of the most important constituents of the ISM. By mass, it makes up only a small fraction of typically about 1%, but dust has important radiative and chemical properties. Dust particles have strong continuum opacities which are highly frequency-dependent. Especially in the optical regime, dust absorbs light much more efficiently than in the infrared regime. That is why young protostars, which are surrounded by gaseous and dusty envelopes, are difficult to observe in the visible wavelengths but require infrared observations. Thermal absorption and reemission of radiation by dust (a process called *reprocession*) strongly determines the thermodynamical properties of a protostellar disc, especially in those regions where the disc is opaque to direct stellar radiation and dominated by thermal reemission of dust molecules. This is mainly the case near the equator of the disc because radial optical depths with respect to the central star are typically much larger than unity ($\tau_* \gg$ 1). Therefore, modeling the temperature structure requires diffuse radiative transfer to be taken into account.

In this test setup, we combine emission from a point source with the solution of the RTE. The goal is to determine the self-consistent temperature structure of the gas in a protostellar disc. The setup is based on the benchmark by Pascucci et al. (2004), which is based on the theoretical work by Chiang and Goldreich (1997). We compare our solutions from a 3D calculation with the results from the Monte Carlo radiative transfer code RADMC-3D (Dullemond, 2012).

Thermal Radiative Transfer

A protostellar disc combines optically thick and thin regimes, which requires the computation of primary stellar radiation and the thermal reemission from dust molecules in the disc. Our approach follows the idea of splitting the radiation field in two components handling each separately. Following the work of Dullemond (2002), the first component we compute is the extinct stellar flux. This can be handled by using the original hybrid characteristics method, which computes the optical depth with respect to a central stellar source (τ_*). The extinct stellar flux F_* at a distance r from a star of luminosity L_* is given by

$$F_*(\mathbf{x}) = \frac{L_*}{4\pi r^2} \exp(-\tau_*(\mathbf{x})),$$

(25)

assuming that the star can be approximated as a point source. The amount of energy per unit time that is absorbed this way is determined by the absorption coefficient χ and given by

$$Q(\mathbf{x}) = \chi F_*(\mathbf{x}).$$

(26)

The reemitted radiation of the dust grains in the disc is treated as a secondary component of the radiation field. This component is computed with the general transfer algorithm using parallel rays. Assuming LTE, the dust grains will acquire an equilibrium temperature such that they emit exactly the same amount of energy which they absorb

$$\frac{\sigma}{\pi} \chi T^4 = \frac{Q}{4\pi} + \chi \frac{1}{4\pi} \oint_{4\pi} I \, d\Omega.$$

(27)

where I is the specific intensity of the reprocessed radiation field. The first term in Eq. (27) accounts for the direct stellar radiation while the second term describes the energy of the reprocessed radiation field. The transfer equation for reemitted radiation by dust grains is

$$\frac{\partial I}{\partial \tau} = \frac{\sigma}{\pi} T^4 - I.$$

(28)

Hence, the source function in this setup is the frequency integrated thermal emission from dust grains $S = \frac{\sigma_{SB}}{\pi} T^4$. The task at hand is to find a temperature that is consistent with the coupled set of Eqs. (27) and (28). This is done by iterating the equations until convergence is reached (Lambda-iteration).

The Disc Model

For the simulation setup we are following the benchmark test of Pascucci et al. (2004)which resembles a *flared disc* (Chiang and Goldreich, 1997). The idea is to define a radial gas surface density distribution and to assume that the vertical density structure is only determined by the hydrostatic equilibrium in the vertical direction. The gas density distribution is given by

$$\rho(r, z) = \rho_0 \, f_1(r) \, f_2(r),$$

$$f_1(r) = \left(\frac{r}{r_d} \right)^{-1.0},$$

$$f_2(r) = \exp\left(-\frac{\pi}{4} \left(\frac{z}{h(r)} \right)^2 \right),$$

$$h(r) = z_d \left(\frac{r}{r_d} \right)^{1.125},$$

(29)

where r is the radial distance in the disc midplane, z is the height above the disc, and ρ_0 is the gas density in the midplane at $r = r_d = 500\,\mathrm{AU}$ and z=0. The outer disc radius is defined by $r_{out} = 1000\,\mathrm{AU} = 2\,r_d$ and we crop the disc at an inner radius=r_{in}. z_d determines the height of the disc which we choose to be $0.25\,r_d$ consistent with Pascucci et al. (2004). We choose the central source to have solar properties with $M_* = 1\,\mathrm{M_\odot}$, $R_* = 1\,\mathrm{R_\odot}$ and $T_* = 5800\,\mathrm{K}$. We use a grey opacity at the visible wavelength of λ=550 nm from the opacity tables used in Pascucci et al. (2004) ($\kappa = 8736\,\mathrm{cm^2\,g^{-1}}$).

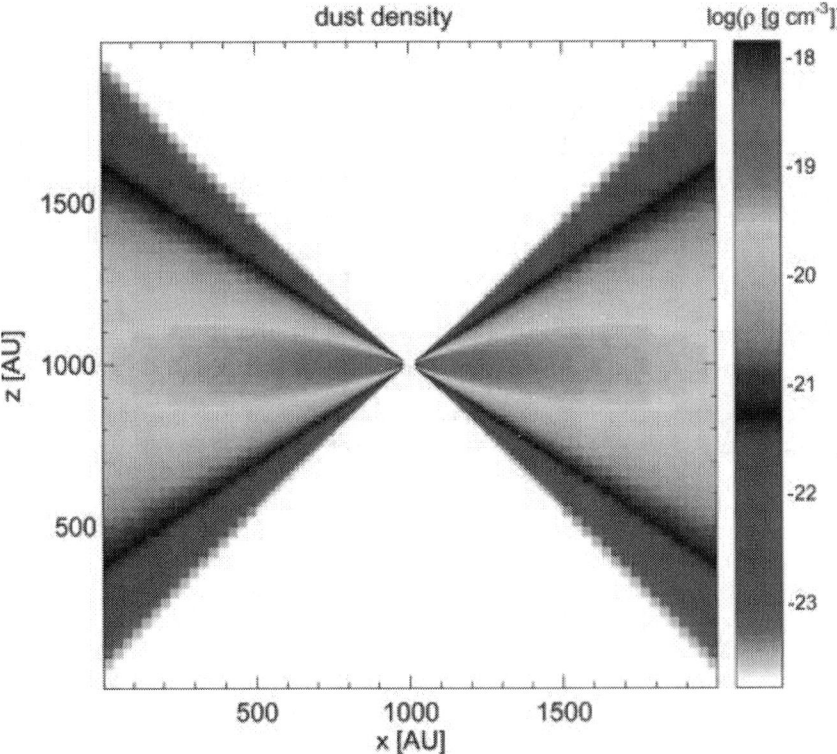

Figure 6. The dust density in the xz -midplane for the Pascucci benchmark for a total optical depth of τ_{disc}=1.

In contrast to the Pascucci benchmark, we perform our calculations in 3D instead of 2D. Therefore, we cannot directly compare our results to the Pascucci results but instead use the results from RADMC-3D as a reference. We perform calculations for three cases of ρ_0 so that the total radial optical depth of the disc in the midplane varies from $\tau_{disc}=1, \tau_{disc}=10$ and $\tau_{disc}=100$. We do not explicitly distinguish between a gas and a dust temperature and assume both to be tightly coupled and the dust density is defined as a fixed fraction of the gas density (1%). The dust density distribution through the xz -midplane of the disc setup for the optically thin case ($\tau_{disc}=1$) is shown in Fig. 6.

The linear spatial resolution varies over 4 refinement levels from $\Delta x = 31.25\,\mathrm{AU}$ in the outer regions to $\Delta x = 1.953\,\mathrm{AU}$ in the center of the disc. The solid angle integration is performed using 768 directions (nSide=8).

Results

The resulting temperature structures and averaged midplane profiles are shown in Fig. 7. As it turns out, the accuracy of the solution is very sensible to the spatial resolution of the inner edge of the disc at $r=r_{in}$ which is a result of discretizing the inner circular rim on a Cartesian grid. Therefore, we increase the inner radius from $r_{in} = 10\,\mathrm{AU}, 20\,\mathrm{AU}$ to 40 AU for the three different setups to guarantee sufficient resolution at the point where the disc becomes optically thick.

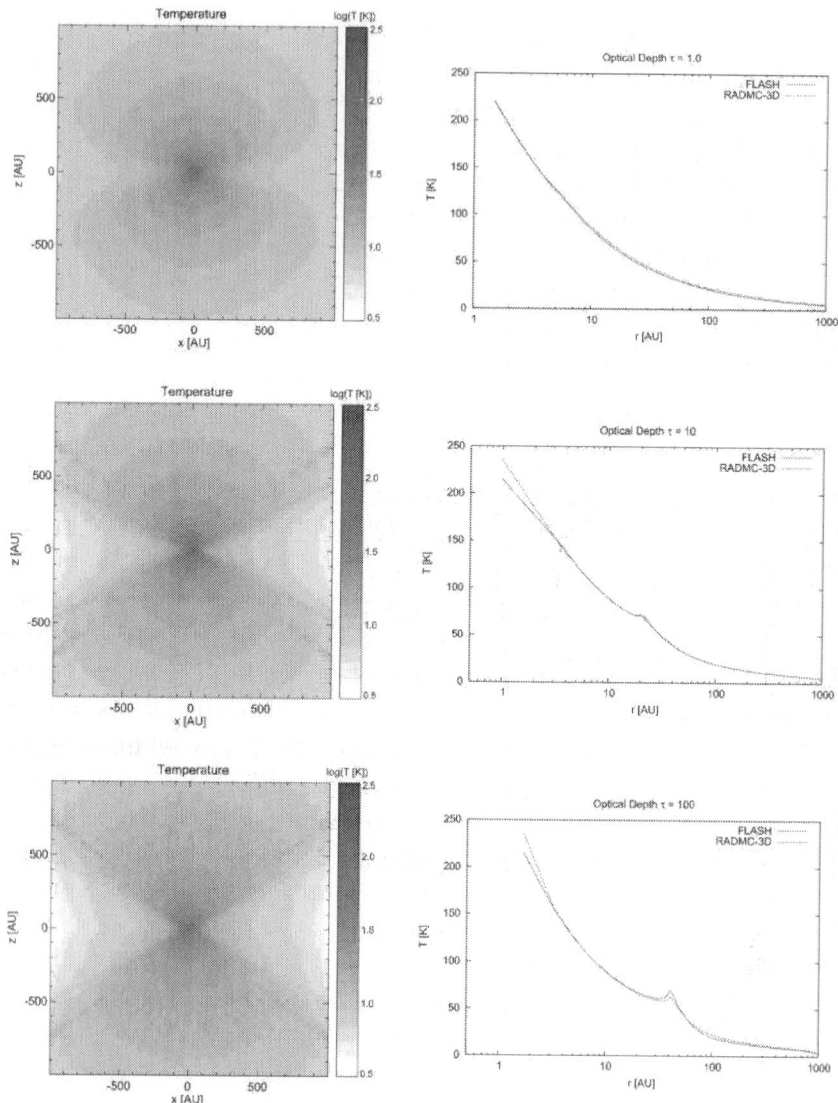

Figure 7. The solutions of the Pascucci et al. (2004) benchmark problem. Left column: the temperature structure through the *xz* -midplane of the disc for total radial optical depths of τ=1 (top), τ=10 (mid), and τ=100(bottom). Right column: averaged temperature profiles in the *xy*-midplane in comparison with the solutions of RADMC-3D. Solutions obtained with FLASH/RT use 768 directions for the angular discretization. Monte Carlo computations with RADMC-3D were performed using 10^8 photon packages.

In the optically thin case ($\tau_{disc}=1$), the midplane temperature is almost entirely dominated by the direct illumination of the central source. In the optically thick cases, the midplane temperature is dominated by the reprocessed radiation from dust in the photosphere of the disc, which is directly illuminated by the central source. At the point where the disc becomes optically thick, a bump emerges in the temperature profile since the dust distribution becomes dense enough to absorb a considerable amount of radiation from the central source. Our results are in excellent agreement with the reference computed by RADMC-3D and within the 10% range of the results from the different codes used for the Pascucci et al. (2004) benchmark.

However, the temperature structure in the left panel of Fig. 7 is sensitive to the angular resolution. Although the raytracer takes care of the known primary stellar radiation, the solution in the outer regions depend on whether the reprocessed radiation from the hot inner rim is accounted for correctly. Especially in the optically thick case ($\tau_{disc}=100$), single rays become visible in the temperature structure even for a large number of directions ($N_{pix}=768$) since not each cell is correctly connected to the hot inner rim in terms of radiative exchange. A larger number of directions is very costly, but an alternative is to also model the emission from such "hot spots" as a part of the primary emission. The problem is to identify these hot spots in the domain since they cannot be represented by point sources or sink particles. But once these regions are identified, their emission can be handled by an inverse raytracer similar to the approach for point sources (Peters et al., 2010). However, this approach requires an adaptive angular grid while our approach is only capable of using a homogenous angular resolution at the moment (Appendix B).

Diffusion Test

In this section, we show results from a time dependent radiative transfer calculation. Solving the time dependent RTE on the timescale of the speed of light would lead to time steps far too small for the use in a hydrodynamical simulation on astrophysical scales. However, since we are not interested in the dynamics of the propagation of the radiation itself but in its contribution to the energy budget of the gas, we assume the hydrodynamical timescale to be much larger than the timescale on which radiation is transported. This means that the radiation field emerges instantaneously everywhere, and we assume the solution of the time independent RTE as being convenient. Consequently, the time dependence of the radiation field originates exclusively from the coupling to the FLASH code using the energy source term from Eq. (8).

In this section, we show results from testing the evolution of the source term by following the propagation of the radiation field in a highly opaque medium. The source function is updated using a simple forward Euler time integration of the energy source term. Since the radiation field shows a diffusion like evolution in the limit of high optical depths, we compare our numerical solution to the analytic solution of the diffusion equation.

Setup

In this test, we investigate the ability of our solver to follow the flux of radiative energy into a highly opaque medium. In this case, the propagation of the radiation field can be described by the diffusion approximation, and we show that our approach reproduces the diffusion limit accurately. The diffusion approximation is derived from the moment equations of the RTE by invoking a closure relation between the radiative energy and the radiative pressure (e.g., the Eddington approximation). The radiation equations themselves then form a hyperbolic system. By neglecting the explicit time dependence of the radiative flux $\mathbf{F}$ and assuming that $\mathbf{F} \propto \nabla E_r$, the flux can be eliminated from the equations. The dynamics of the radiation field $J = c E_r / (4\pi)$ can then be described in a single equation, the diffusion equation (Mihalas and Weibel Mihalas, 1984):

$$\frac{\partial J}{\partial t} - \nabla\left(\frac{c}{n\,\chi}\nabla J\right) = c\,\chi\,(S - J).$$

(30)

where n denotes the number of dimension. We do not allow any interaction of the radiation field with the hydrodynamics and only follow the propagation of the radiation field. Hence, the diffusion equation becomes homogeneous since $S = J$. In this case, the solution to the diffusion equation is described by the Gaussian function

$$J_D(\mathbf{x}, t) = \frac{J_0}{(4\pi Dt)^{n/2}} \exp\left(-\frac{(\mathbf{x} - \mathbf{x}_0)^2}{4Dt}\right),$$

(31)

where J_0 denotes the initial mean intensity at $t = t_0$, $\mathbf{x}_0$ is its initial position, and $D = c/(\eta\chi)$ is the diffusion coefficient. We use Eq. (31) to compute the initial conditions $J(\mathbf{x}, t_0)$ for our test setup. We perform 1D and 3D computations with the initial conditions $J_0 = J(\mathbf{x}_0, t_0) = 10^5 \, \text{erg s}^{-1} \, \text{cm}^{-2} \, \text{sr}^{-1}$ with $t_0 = 10^{-11} \, \text{s}$ in 3D and $t_0 = 10^{-10} \, \text{s}$ in 1D respectively, The center of the Gaussian is at $\mathbf{x}_0 = 0$, and we evolve the radiation field until $t = 20 \times t_0$ is reached. The length of the computational domain is 1 cm with a homogeneous density distribution of $\rho = 1 \, \text{g cm}^{-1}$ and a constant absorption coefficient $\kappa = 1000 \, \text{cm}^2 \, \text{g}^{-1}$, which results in a highly optically thick medium. The temperature is constant and arbitrarily set to $T = 1 \, \text{K}$. Since no heating or cooling is allowed, there is no hydrodynamical response from the medium and all hydrodynamical quantities are constant in space and time.

Since we solve the time-independent RTE, there is a problem in reproducing the time-dependent term in Eq. (30). Strictly speaking, the static source function vanishes since we do not couple the radiation field to the medium through which it propagates. Consequently, the mean intensity would also vanish in the time independent solution. However, the time dependence causes an effective contribution in the source function (e.g. Jack et al., 2012) if the time discretization is carried out implicitly in the RTE (1). This contribution depends on the specific intensities of the previous time step, is evolved through time, and describes the evolution of the radiation field. Since we do not account for this implicit contribution (which would require to store the complete scalar field of angle dependent specific intensities), we solve the problem by operator splitting using the right-hand side of Eq. (30) to calculate the new source function at the following time step. The evolution is done using a simple forward Euler time integration scheme of the form

$$S_n = S_{n-1} + \Delta t_n \, \chi \, c \, (J_{n-1} - S_{n-1}),$$

(32)

where Δt_n is the length of the current time step n. Therefore, the time step is restricted to be (Section 2.4.1)

$$\Delta t_n = \min \left(\frac{S_{n-1}}{(|S_{n-1} - S_{n-2}|)} \right)_i k_{\text{rad}} \Delta t_{n-1},$$

(33)

where the min function denotes the minimum change in the source function with time from all cells i in the computational domain (FLASH does not support adaptive time stepping on a block level but rather uses a uniform global time step). k_{rad} limits the maximum change in the source function, and we found a value of $k_{\text{rad}} \approx 0.1$ to give stable and accurate results in 3D.

Results

The results of the 1D solutions are shown in Fig. 8. We compare the numerical results with the analytic solution given by Eq. (31) and found our results to be within 1% accuracy at a resolution larger than 32 cells. At the edge of the domain, the numerical solution deviates from the diffusion solution as radiative energy can leave the domain and we allow no irradiation from the outside. The results from the 3D computation are shown inFig. 9 and compared to the diffusion solution along the three main axes of the domain. In the 3D case, the domain is subdivided by the AMR grid into 4 blocks in each dimension. Each block contains 8^3 cells giving a total linear resolution of 32 cells. In the 3D case, the setup consists of a Gaussian kernel around the origin which diffuses outwards. The solutions along each coordinate axis are obviously indistinguishable, emphasizing the accuracy and importance of the homogeneous angular HEALPix tessellation (Appendix B). The 3D computations were performed using 192 directions.

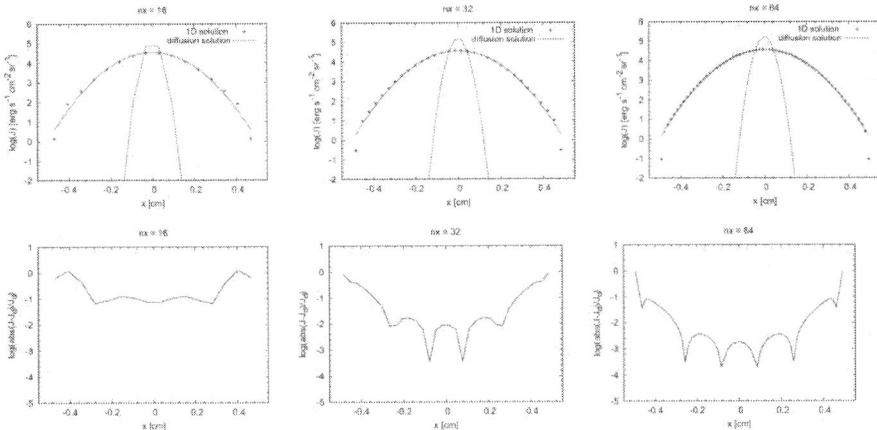

Figure 8. Results of the 1D diffusion test for different homogeneous spatial resolutions, *top:* nx = 16, *mid:* nx = 32,*right:* nx = 64. The dashed lines show the initial conditions at $t = t_0$ determined by the Gaussian solution of the diffusion equation. The initial radiative energy (symbols) is evolved and diffuses outwards until $t = 20 \times t_0$ is reached and compared to the analytical solution (solid lines) of the homogeneous diffusion equation. For a sufficient spatial resolution, the numerical solution stays within 1% accuracy. At the edge of the domain, the radiation solver deviates from the diffusion solution as radiation leaves the domain while the diffusion solution is valid for an infinite domain.

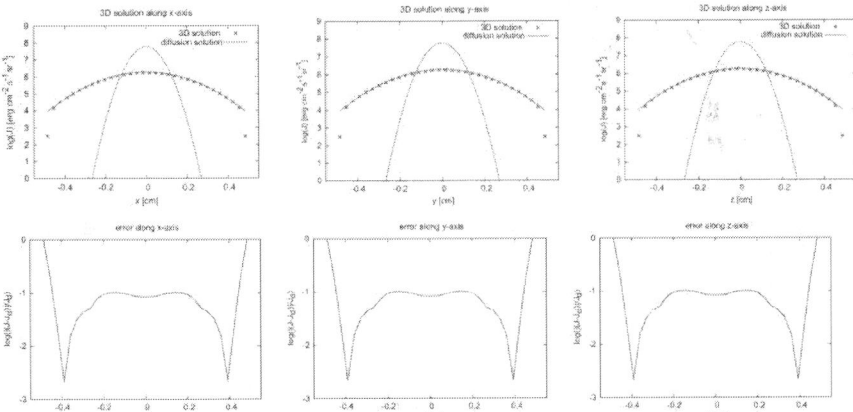

Figure 9. Results of the 3D diffusion test along the *x*-(*left*), *y*-(*mid*) and *z*-axis (*right*) of the simulation box with a homogeneous spatial resolution of *nx=ny=nz=*32 (symbols). *Top row:* The dashed lines show the initial conditions at $t = t_0$ determined by the Gaussian solution of the diffusion equation. Blue solid lines show the analytic solution according to Eq. 31. Symbols show the numerical results from our radiation solver.*Bottom row:* The relative error of the numerical solution. The 3D solution is not as accurate as the 1D results but still within 10% of the analytical solution. The obvious independence of the solution on the direction axis results from the homogeneous angular HEALPix tessellation. The calculations were performed using 192 directions. (For interpretation of the references to color in this figure legend, the reader is referred to the web version of this article).

1D Non-equilibrium Radiative Shocks

Testing the radiative transfer solver for radiative shock computations is the next crucial step and requires the combination of our radiation solver with the FLASH code. The source term is determined by the energy budget of absorption and emission processes. We recall the frequency integrated source term from Eq. (8) here:

$$Q_{rad} = 4\pi\chi_a(J - B), \tag{34}$$

which is coupled to the hydrodynamical solver by adding it to the right-hand side of the Euler equation for the internal gas energy. For this test case, the emission and absorption opacities are equal. Since the shock setup is used for test purposes, we neglect the magnetic field.

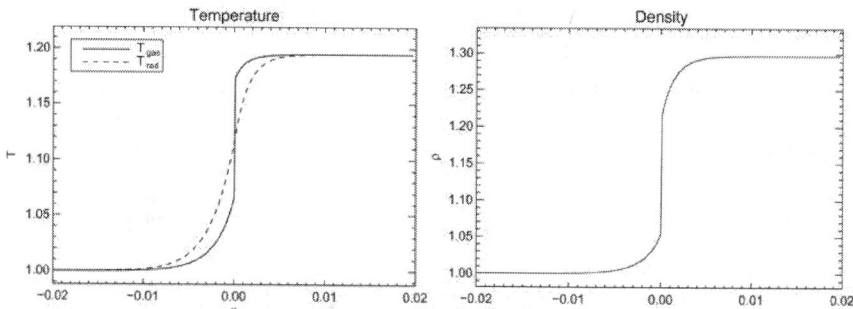

Figure 10. Normalized temperature and density profiles for the subcritical shock with $M_0 = 1.2$ in the equilibrium state after 10 ns. The gas is preheated on the upstream side and cools down on the downstream side of the hydro shock front.

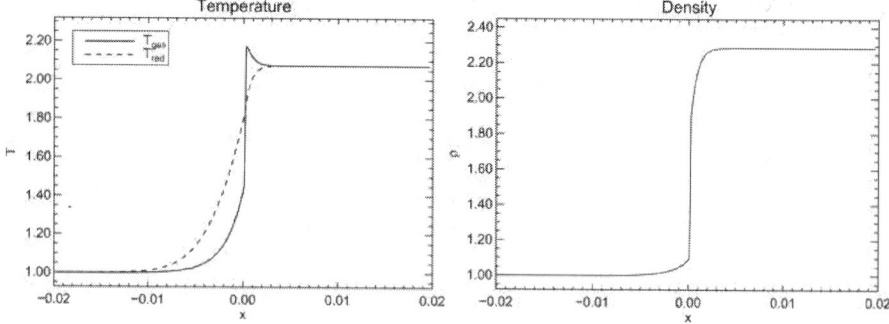

Figure 11. Same conditions as in Fig. 10 but with $M_0 = 2$. The maximum temperature at the shock begins to exceed the downstream equilibrium temperature which results in the Zel'dovich spike. Since the temperature at the upstream side of the shock is still well below the downstream temperature, the shock is subcritical.

Initial Conditions

The initial conditions are consistent with the theoretical work of Lowrie and Edwards (2008). In their work, the jump conditions and the equations of radiation hydrodynamics are given in a dimensionless form. The equations are normalized using reference material quantities and a constant P_0 which arises from the normalization process and is given by

$$P_0 = \frac{\tilde{\alpha}_r \tilde{T}_0^4}{\tilde{\rho}_0 \tilde{a}_0^2}.$$
(35)

The quantities denoted with a tilde are the dimensional reference material attributes (temperature $\tilde{T}_0$, density $\tilde{\rho}_0$, sound speed $\tilde{a}_0$) and $\tilde{\alpha}_r$ is the radiation constant. The "0"-subscript indicates pre-shock state initial values. P_0 gives a measure for the relative importance of gas and radiation pressure or alternatively, the radiative energy to the material energy (Mihalas and Weibel Mihalas, 1984). For our test setups, we choose a grey non-equilibrium shock setup with Mach numbers of $M_0=1.2, M_0=2$ (subcritical), and $M_0=5$ (supercritical), which we compute in the reference frame of the shock with $P_0 = 10^{-4}$ and $\gamma = 5/3$. Lowrie and Edwards (2008) give a dimensionless absorption and transmission cross section, which determine the radiative energy exchange and diffusivity of the radiating materials. Evaluating the dimensionless values gives an absorption coefficient of $\kappa_a \approx 423.0\,\mathrm{cm^2/g}$ and a total extinction coefficient of $\chi \approx 788.0\,\mathrm{cm^2/g}$, which results in an effective photon destruction probability of $\epsilon = \kappa_a/\chi \approx 0.5377$. The initial dimensionless pre-shock gas temperature T_0 and density ρ_0 are set to unity, the post-shock initial values (T_1, ρ_1) are computed using the Rankine–Hugoniot jump conditions. The actual dimensional initial conditions can then be calculated using their dimensional reference material values (for more details we refer to Lowrie and Edwards (2008)). Finally, the radiation temperature

$$T_r = \left(\frac{\pi}{\sigma_{\mathrm{SB}}} J \right)^{1/4}$$
(36)

is initially in equilibrium with the gas temperature. For the radiation shock test problem, the source function is determined by a thermal emission and a diffusive part. This is equivalent to using the isotropic scattering source function

$$S = (1-\epsilon)J + \epsilon B \tag{37}$$

with the appropriate photon destruction probability and a thermal energy contribution given by the frequency integrated Planck emission $B = \frac{\sigma_{SB}}{\pi}\tilde{T}^4$. Since the radiation field will not be not be in thermal equilibrium with the material throughout the simulation, we need to iterate until a consistent solution of the mean intensity J is found. However, since $\epsilon \approx 0.5377$ gives only a moderate scattering contribution and using the solution from the previous time step, the accelerated lambda iteration usually converges after 2 or 3 iteration steps.

Results

The shocks need a few nanoseconds to relax into a static equilibrium state. Figs 10, 11and 12 show the resulting temperature and density profiles after 10 nanoseconds. Sufficiently far upstream (left) and downstream (right) of the hydrodynamical shock (atx=0), gas and radiation are in a thermodynamical equilibrium and the radiation temperature coincides with the gas temperature computed from the initial conditions. Since the total extinction coefficient χ is about twice the thermal absorption and emission coefficient, the temperature of the radiation field and the gas are out of equilibrium near the shock front.

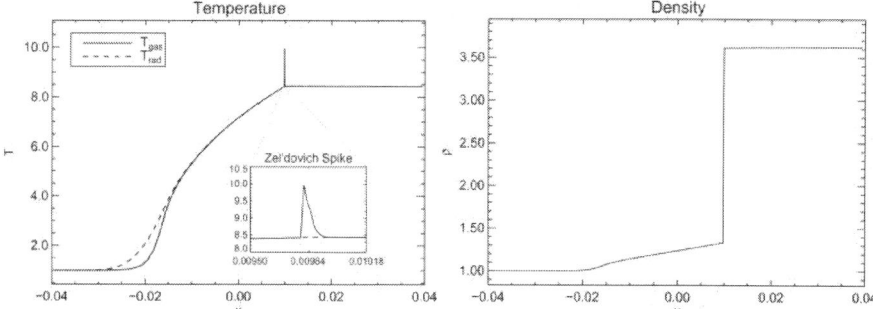

Figure 12. Same conditions as in Fig. 10 but with $M_0 = 5$. The temperature on the upstream side of the hydro shock front reaches the downstream equilibrium value. The Zel'dovich spike gets very narrow and the shock becomes supercritical.

The subcritical shock with $M_0 = 1.2$ (Fig. 10) shows a hydro shock but no spike in the radiation temperature. For $M_0 = 2$ the so called *Zel'Dovich spike* in the gas temperature appears for the first time as seen in Fig. 11. The spike appears since radiation is transported through the hydrodynamical shock and preheats the inflowing gas, which is initially in a thermal equilibrium with the radiation field in the upstream region. After the gas has passed the hydrodynamical shock, it cools down until the radiation field and the gas are again in thermal equilibrium on the downstream side of the shock. Since the upstream temperature at the shock front is still less than the downstream temperature the shock is subcritical. For $M_0 = 5$, the shock becomes supercritical, since the upstream gas is preheated until it reaches the downstream gas temperature even before passing the hydrodynamical shock front. The discontinuity in the gas temperature is then restricted to the narrow range of the Zel'dovich spike (Fig. 12). Our solutions resemble the semi-analytical results from Lowrie and Edwards (2008) and show the correct spike evolution. However, a closer look at the results show a slight deviation of the shock front from its initial position (at $x = 0$). Especially in the supercritical case, the shock front drifts very slowly into the downstream direction. This drift is due to the absence of the radiation pressure in our approach, which becomes important for high Mach numbers (with a high downstream gas temperature). While the shock front drifts very slowly, the temperature and density profiles do not change since the radiation source term is still very well approximated in our approach.

3D COLLAPSE SIMULATIONS

In this section, we show results from full 3D radiation hydrodynamical simulations performed with FLASH/RT. Since we aim to use our framework for the modeling of radiative feedback in star formation simulations, we show the capabilities of our method in two self-gravitating collapsing cloud simulations. We follow the collapse until the first hydrostatic core is formed and before the dissociation of hydrogen molecules start (the first collapse). In Section 4.2, we show results from a basic collapse simulation without rotation and compare the resulting profiles to other similar works. Afterwards, we show results from a more complex simulation including rotation and turbulence (Section 4.3) and compare the results to a simulation without modeling radiative transfer. The angular resolution of the radiative transfer calculations are the same for both collapse simulations, and we use 768 directions to compute the radiation field (nSide = 8 for the HEALPix tessellation).

Opacities

Since our solver does not yet support any frequency dependence, the source function S is only determined by the frequency-integrated thermal emission of the gas ($S = B = \frac{\sigma_{SB} T^4}{\pi}$), and we neglect any scattering processes. Consequently, we have to use frequency-integrated mean dust opacities. For this purpose, we choose the Planck mean opacities by Semenov et al. (2003). In their work, the dust composition model takes into account the evaporation temperatures of ice, silicates, iron as well as their density dependencies. We coupled their subroutines [3]for computing temperature and density dependent dust opacities into FLASH, and we choose the input parameters for spherical homogeneous dust grains with a normal relative iron content in the silicates of Fe/(Fe+Mg) = 0.3.

Collapse without Rotation

In this section, we study the collapse of a spherical, homogeneous, and gravitationally unstable density distribution. The initial conditions do not contain any turbulence or density perturbations and hence, the results are spherically symmetric. This setup represents a common benchmark for the capabilities of a radiation hydrodynamical astrophysical computer code, and we compare our results to similar work done byCommerçon et al. (2011), Masunaga et al. (1998), and the pioneering simulations ofLarson (1969).

Initial Conditions

We start with highly gravitationally unstable initial conditions. The cloud core of one solar mass consists of a homogeneous sphere with radius $R_0 = 7.07 \times 10^{16}\,\text{cm}\ (\approx 4725\,\text{AU})$ and and density $\rho_0 = 1.38 \times 10^{-18}\,\text{g cm}^{-3}$, which results in an initial free fall time of $t_\text{ff} \approx 56.67$ kyrs. The linear size of the 3D computational domain is four times the initial cloud radius R_0 in each dimension. The surrounding gas density is a hundred times less than the initial cloud density ρ_0, and the cloud is initially in thermal equilibrium with the ambient gas at a temperature of $T_0 = 10\,\text{K}$ resulting in an initial isothermal sound speed of $c_s \approx 0.195\,\text{km s}^{-1}$. Since the cloud is initially not in pressure equilibrium with its surroundings, FLASH's hydrodynamical solver drives a weak shock wave into the ambient gas which is soon dissipated. To prevent our radiation solver from resolving this shock in terms of radiative energy exchange (which would result in rather small time steps), we do not couple the radiation field to the hydrodynamics outside of R_0 but rather keep the ambient gas and radiation temperature fixed.

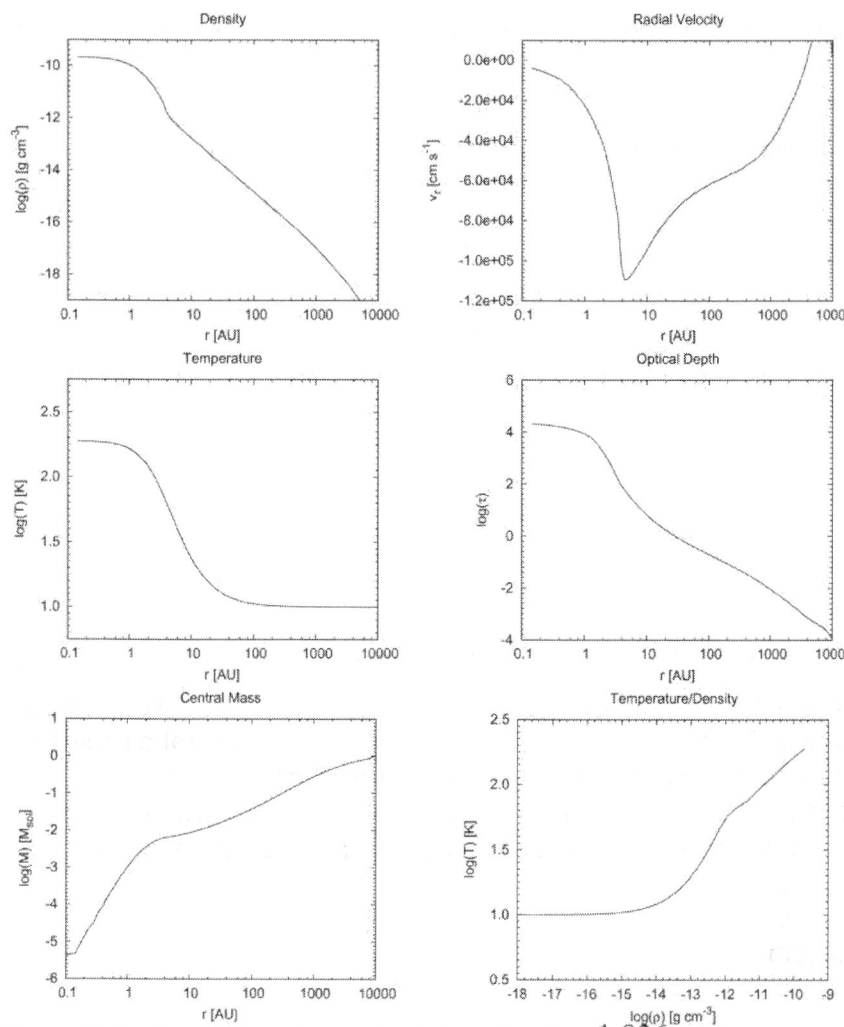

Figure 13. Profiles of the collapse simulation after $t = 1.036\,t_{\mathrm{ff}}$; the maximum density at the core center is $\rho_c \approx 2.0 \times 10^{-10}\ \mathrm{g\,cm^{-3}}$ with a temperature of $T_c \approx 186\ \mathrm{K}$, a radius of $R_{\mathrm{fc}} \approx 4\ \mathrm{AU}$ and a mass of $M_c \approx 10^{-2}\ \mathrm{M_\odot}$.

The initial conditions result in a gravitationally unstable cloud core which contains nearly two Jeans masses. To ensure a proper resolution and avoid artificial fragmentation during the collapse, we use the Jeans condition by Truelove et al. (1997) as the refinement criterion of the AMR grid. In our case, we use at least $N_j = 9$ grid cells per Jeans length. To resolve the first hydrostatic core properly, we allow a maximum linear resolution of $\Delta x \approx 0.07$ AU which requires the AMR grid to cover 11 levels of resolution.

The summarized initial conditions are:

$$\text{Mass} \quad M = 1.0 \, M_\odot,$$
$$\text{Density} \quad \rho_0 = 1.38 \times 10^{-18} \, \text{g cm}^{-3},$$
$$\text{Temperature} \quad T_0 = 10 \, \text{K},$$
$$\text{Angular velocity} \quad \Omega = 0.0 \, \text{rad s}^{-1},$$
$$\text{Radius} \quad R_0 = 7.07 \times 10^{16} \, \text{cm},$$
$$\text{Free fall time} \quad t_{\text{ff}} = 56.67 \, \text{kyrs}.$$

Results

The cloud core starts to collapse, and as soon as the maximum density in the cloud exceeds about $10^{-13} \, \text{g cm}^{-3}$, the central regions of the cloud core become optically thick. At this point, the central temperature starts to rise rapidly and the following evolution proceeds almost adiabatically with more gas falling onto the central quasi-hydrostatic core. Since the simulation does not contain any rotation or turbulence, the 3D solution is spherically symmetric, and we present the results in the form of averaged radial profiles. The profiles for density, radial velocity, temperature, optical depth, and central mass after $1.036 \times t_{\text{ff}}$ are shown in Fig. 13. The resulting protostellar core has a mass of $M_{\text{fc}} \approx 1 \times 10^{-2} \, M_\odot$, a radius of $R_{\text{fc}} \approx 4$ AU, and a central temperature of $T_c \approx 186$ K. The boundary of the core can be identified easily in the velocity profile, where there is a sudden decrease in the infall velocity (the accretion shock). Inside the core, the infall does not stop completely indicating that the core is only quasi-hydrostatic.

Table 1. Comparison of simulation results; R_{fc} is the radius of the first core, M_{fc} is the core mass, T_{fc} the central temperature and T_{fc} is the temperature at R_{fc}.

Reference	R_{fc} [AU]	$M_{fc}[M_\odot]$	$T_{fc}[K]$	T_c [K]
This work	4	1×10^{-2}	50	186
Commerçon et al. (2011)	8	2.1×10^{-2}	81	396
Masunaga et al. (1998)	8	$\approx 10^{-2}$	60	200
Larson (1969)	4	1×10^{-2}	–	170

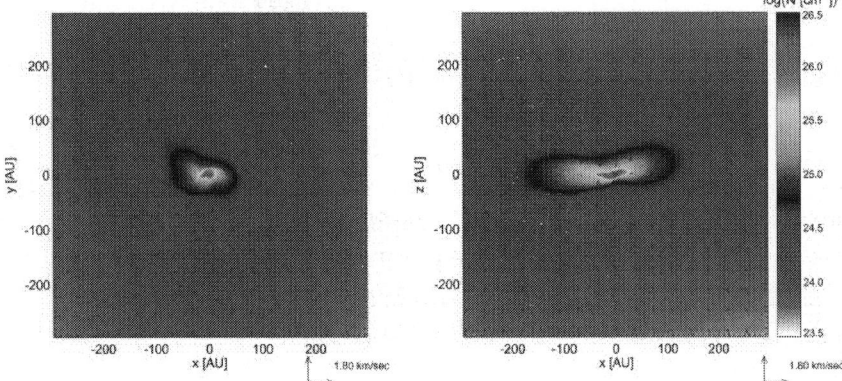

Figure 14. Column densities along the z- (left) and y-axis (right) of the simulation box after the formation of the first hydrostatic core at $t \approx 60\,$kyrs $\approx 1.07\,t_{ff}$ including rotation and turbulence. The rotational energy forces the gas to accumulate in a circumstellar disc (in the xy-plane) around the first core.

Our results are quantitatively very similar to those of Larson (1969) and qualitatively very similar to the more recent works by Masunaga et al. (1998) and Commerçon et al. (2011).Table 1 shows an overview of the characteristic temperature, mass and radius of the first core in comparison to these works (the common reference point is when the maximum central density of the first core reaches $\rho_{fc} \approx 2 \times 10^{-10}\,\mathrm{g\,cm^{-3}}$). Apparently, our computations produce qualitatively similar results, although the methods invoked in the other works are quite different and use different initial conditions and opacity models.

Collapse with Rotation and Turbulence

This simulation run has very similar initial conditions as described in the previous section except that we add rotational and turbulent energy. The cloud is initially in a rigid body rotation around the z-axis at the center of the simulation box. The ratio of rotational and gravitational energy is given by

$$\beta = \frac{1}{3} \frac{R_0^3 \Omega_0^2}{G M_0}.$$
(38)

We choose $\beta = 0.03$ which gives an initial angular velocity of $\Omega_0 = 1.886 \times 10^{-13} \text{ rad s}^{-1}$ and agrees with typically observed values of molecular cloud cores (Goodman et al., 1993). In addition, we superimpose a turbulent velocity perturbation on the initial uniform angular velocity field. The construction of the velocity perturbation is based on the theory for incompressible turbulence by Kolmogorov (1941), in which the kinetic energy E of the velocity fluctuation with wave number k is described by a power spectrum

$$E(k) \propto k p.$$
(39)

The wave number $k = 2\pi/l$ is the inverse of the length scale l of a turbulent fluctuation (sometimes called *eddy*). In our case, the spectrum has a power law index of $p = -2$ resembling a Burgers type model of turbulent energy decay. The geometries and density distribution of the initial cloud core are the same as for the simulation without rotation and turbulence.

In addition to the simulation run with FLASH/RT, we also run the simulation without modeling radiative transfer. Instead, we use a barotropic EOS with a density-dependent effective adiabatic exponent γ that mimics radiative cooling. The internal energy/temperature is fixed at

$T_0 = 10\,\mathrm{K}$ as long as the gas density is less than $\rho \approx 10^{-15}\,\mathrm{g\,cm^{-3}}$ (isothermal). Above this threshold density, the temperature rises slowly with $\gamma = 1.1$ until the adiabatic exponent becomes $\gamma = 4/3$ above $\rho \approx 10^{-13}\,\mathrm{g\,cm^{-3}}$ (adiabatic). We ran the simulation including radiative transfer as well as the reference run with the barotropic EOS until the formation of the first hydrostatic core with a central density of $\rho_{\mathrm{fc}} \approx 10^{-11}\,\mathrm{g\,cm^{-3}}$. At this point, both simulations cover 9 different levels of resolution in the AMR grid with a maximum linear resolution of $\Delta x \approx 0.57\,\mathrm{AU}$ while the whole simulation box has a linear extent of 18, 903 AU.

The summarized initial conditions are:

$$
\begin{aligned}
\text{Mass} \quad & M = 1.0\,M_\odot, \\
\text{Density} \quad & \rho_0 = 1.38 \times 10^{-18}\,\mathrm{g\,cm^{-3}}, \\
\text{Temperature} \quad & T_0 = 10\,\mathrm{K}, \\
\text{Angular velocity} \quad & \Omega = 1.886 \times 10^{-13}\,\mathrm{rad\,s^{-1}}, \\
\frac{\text{Rotational energy}}{\text{Gravitational energy}} \quad & \beta = 0.03, \\
\text{Radius} \quad & R = 7.07 \times 10^{16}\,\mathrm{cm}, \\
\text{Free fall time} \quad & t_{\mathrm{ff}} = 56.67\,\mathrm{kyrs}.
\end{aligned}
$$

Results

The rotational energy and the superimposed turbulent velocity perturbations break the symmetry of the simulation. Fig. 14 shows the column densities along the main axes of the inner region where the dense first core has formed after about 60 kyrs ($\approx 1.07\,t_{\mathrm{ff}}$) with a maximum gas density of $\rho_{\mathrm{fc}} \approx 10^{-11}\,\mathrm{g\,cm^{-3}}$. Because of the additional rotational and turbulent energy, the formation of the first core is deferred and forms later than in the previous simulation (Section 4.2). The conservation of angular momentum causes the first core to be flattened roughly along the z-axis and the density distribution shows a flat disc-like structure revolving around the central compact hydrostatic core. The resulting density distribution is roughly the same as in the reference run without radiative transfer. The initial collapse which seeds the formation of the central core does mostly occur in the isothermal phase, hence, modeling radiative feedback does not influence the initial formation of the core significantly. However, Fig. 15 shows the resulting density weighted temperature averages along the main axes in the central regions around the first core (e.g. $\int \rho\, T\, dz / \int \rho\, dz$). The left column shows the results including radiative transfer (FLASH/RT) while the right column shows results from the reference run. The FLASH/RT model clearly shows how the central core heats the surrounding gas to a temperature roughly 30% higher than in the reference run (like in Price and Bate, 2010). The resulting temperature density distribution in comparison to the barotropic EOS is shown in Fig. 16.

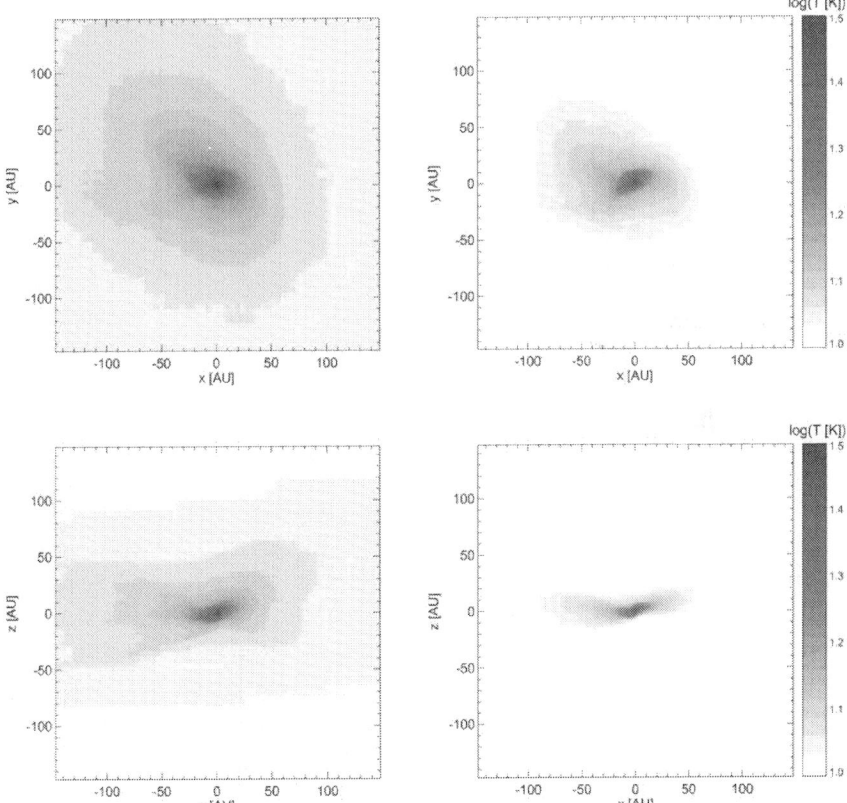

Figure 15. The plots show density weighted temperature averages (e.g., $\int \rho\, T\, dz / \int \rho\, dz$) from a collapse calculation including rotation and turbulence. *Left:* results from the FLASH/RT calculations including radiative transfer. *Right:* results from FLASH calculations using a barotropic EOS. The ambient gas temperature in the FLASH/RT models is about 30% higher.

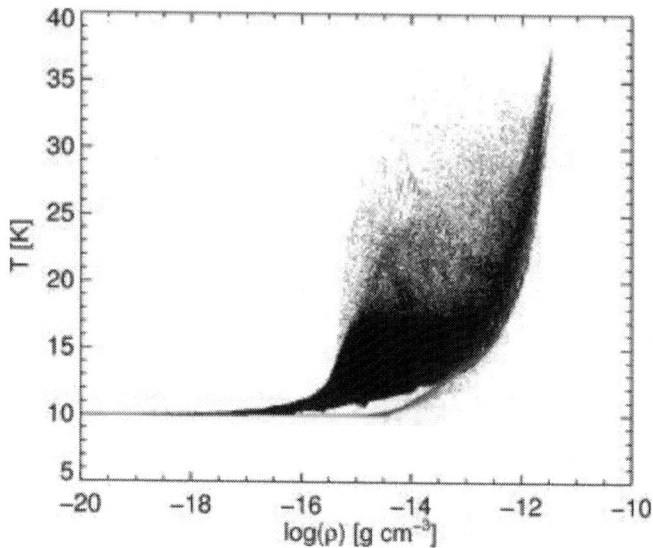

Figure 16. Temperature distribution with respect to the gas density in the simulation box at the end of the collapse simulation including rotation and turbulence. Black dots show the temperature distribution from the FLASH/RT run, red dots resemble the temperature density dependence of the barotropic EOS. (For interpretation of the references to color in this figure legend, the reader is referred to the web version of this article).

Unfortunately, our FLASH/RT simulations are very costly (see Section 5 for more details) and currently, it is not feasible to continue these simulations without coupling the radiative transfer solver to a subgrid model for the formation of the central core, e.g., sink particles (Federrath et al., 2010). However, our current test simulations show the first stages of disc formation and the importance of modeling radiative transfer accurately. Since the thermodynamics of the gas significantly influence the fragmentation behavior, modeling radiative transfer is indispensable to study the further evolution of the protostar, the circumstellar disc, and the surrounding gas envelope.

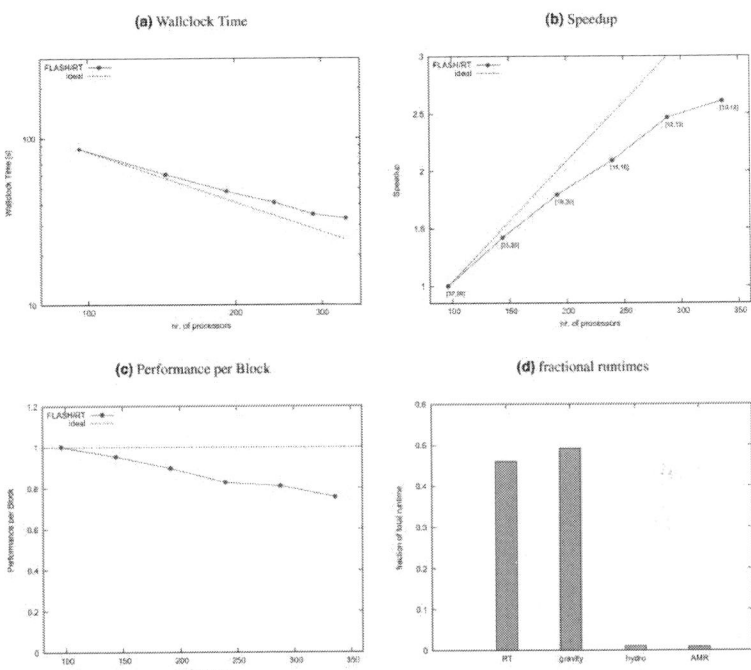

Figure 17. Results from the parallel scaling test. (a) The total wall-clock time for the formal solution averaged over 50 iteration cycles from the Pascucci disc benchmark is shown. (b) The speedup normalized to the wall-clock time using 96 cores is shown. The number in brackets denote the minimum and maximum blocks per core, which is the result from the Morton space-filling curve. (c) The performance per block is shown, which decreases by roughly 10% if the number of cores is doubled. (d) The fractional runtimes for the most costly steps of the collapse simulation (Section 4.2) is shown. All computations were performed using 192 directions.

Table 2. Results from the scaling test normalized to a run with 96 cpus'; because of the increased communication overhead, each cpu should handle as many block as possible in terms of memory requirements.

No. of cores	Time [s]	Speedup	Blocks per cpu	Performace per Block [%]
96	86.06	1	37–39	100
144	60.6	1.42	25–26	95.2
192	48.01	1.79	18–20	89.6
240	41.11	2.09	14–16	82.6
288	34.93	2.46	12–13	81
336	32.98	2.6	10–12	75.5

PERFORMANCE

The FLASH code shows excellent scaling behavior on any computational infrastructure (e.g. Fryxell et al., 2000). For this work, the computations are clearly dominated by the solution of the RTE. Hence, the scaling behavior of the radiative transfer solver is crucial for the total performance of the FLASH/RT calculations. We investigate the scaling performance of our radiation code using the disc benchmark setup (see Section 3.2). We performed 50 formal solutions of the RTE using 192 directions on a spatial range covering 5 refinement levels. After the initial refinement, depending on the density structure and the radius, the computational domain consists of 3648 valid subdomains (leaf blocks) each containing 8^3 cells. The FLASH code distributes the blocks among all available MPI ranks using a Morton space-filling curve[4]. The scaling tests were run at The North-German Supercomputing Alliance in Berlin on the Cray XC30 "Gottfried" using 12-core Xeon IvyBridge processors. Fig. 17 and Table 2 show the scaling results for the computation of the formal solution of the RTE averaged over 50 cycles. The scaling is normalized to the wall-clock time using 96 cores (e.g., 8 Xeon IvyBridge processors). "Gottfried" provides 2 Xeon processors with 24 cores in total per computing node, hence, adding 24 cores to the computation will increase the communication overhead. Fig. 17b shows the speedup compared to a perfect scaling behavior. The radiation solver scales reasonably well considering the communication of non-local information, which is necessary for the solution of the RTE. Fig. 17c shows that doubling the number of cores decreases the performance per block by approximately 10%, which we consider also as reasonable.

The cost of the radiative transfer solver from a 3D collapse simulation (Fig. 17d) is comparable to the cost for the computation of the self-gravitational potential which is done by a Poisson tree-solver. However, the radiative transfer solver in this particular simulation uses a rather moderate angular resolution of 192 directions (using the HEALPix tessellation from Górski et al., 2005). For runs including rotation and turbulence, the angular resolution probably needs a much higher resolution of at least 768 directions or higher. Since the cost of the radiative transfer solver scales linearly with the number of directions, it dominates the entire simulation run compared to the calculation of self-gravity. So far, we have tested the FLASH/RT code on our own computing cluster in Hamburg (32 nodes with 2x Intel Xeon Hexa-Core CPUs, 2.40

GHz) and at the North-German Supercomputing Alliance in Berlin on the Cray XC30.

SUMMARY

We have implemented a new radiation transfer solver based on the method of hybrid characteristics. The solver successfully reproduces standard radiative transfer problems, including NLTE, thermal radiative transfer and the diffusion limit. We proved the feasibility of the method for 3D collapse simulations where radiative transfer is the dominant cooling process during the formation of the first protostellar core. In contrast to the FLD approximation, our method preserves the anisotropy of the radiation field which becomes crucial in the transition from optically thin to optically thick regions (e.g., a protostellar disc). The radiation solver is implemented in the framework of the MHD code FLASH which allows for a straight forward coupling of both codes (e.g., the collapse simulations.

However, the explicit energy coupling, as described in Section 2.4.1, puts rather strong limitations on the time step. A possible improvement can be achieved by combining the raytracer with the solution of the moment equation for the radiative energy. In contrast to the FLD approach, one can compute an angle dependent diffusion coefficient in the form of the variable Eddington tensor (VET) (Jiang et al., 2012) which can be achieved using our raytracer. The advantage is that the evolution of the radiative energy can be handled implicitly by solving the linearized moment equation for the radiation temperature, like inCommerçon et al. (2011), which resolves the problem of the time step restriction. The framework for this has already been implemented in FLASH by Klassen et al. (2014) and can be combined with our raytracer to implement the VET approach.

Our method is a generalized and enhanced implementation of the hybrid characteristics method by Rijkhorst et al. (2006) and Peters et al. (2010). The original implementation was restricted to direct irradiation from point sources and the integration of optical depths respectively. The FLASH code in combination with our radiative transfer framework allows for the solution of a much wider range of problems and can also very easily be extended to handle a more complex form of the radiative transfer equation. Our implementation fits very well into the parallel design of the FLASH code which is based on AMR with domain decomposition. Our method works within the AMR design of FLASH and is able to solve the 3D RTE on a wide range of scales which is indispensable for star formation simu-lations.

ACKNOWLEDGMENTS

L.B. acknowledges financial support by the Deutsche Forschungsgemeinschaft (DFG) mainly via the Emmy Noether Grant BA 3706/1-1*Theory of Massive Star Formation* and partly through the Graduiertenkolleg 1351 *Extrasolar Planets and their Host Stars* and the Priority Program 1573 *Physics of the Interstellar Medium*. T.P. acknowledges financial support through a Forschungskredit of the University of Zurich, Grant no. FK-13-112, and from the DFG Priority Program 1573 *Physics of the Interstellar Medium*. This work benefited from helpful discussions with Peter Hauschildt (Universitä Hamburg) and Stefan Dreizler (Universität Göttingen). Most of the collapse simulations were carried out at The North-German Supercomputing Alliance in Berlin (*Gottfried*, HLRN).

APPENDIX A. ACCELERATED LAMBDA ITERATION

The lambda operator Λ describes the task to compute the radiation field from the source function. It is usually written as

$$J = \Lambda[S]. \tag{A.1}$$

Formally, we can solve this by inverting the Lambda operator. When we arrange the cells of a 3D domain successively in a 1D vector, we can write the operator as a matrix. But the complete operator for *one cell* in the computational domain contains all radiative contributions from each other cell. Hence, the Lambda matrix is far from being sparse. The explicit construction and storage of the Lambda matrix would easily reach computational limits in terms of memory requirements. Furthermore, the inversion of the Lambda operator is far too costly to be used in 3D radiative transfer. Instead, the formal solution (4) is used. Since the source function may depend on the mean intensity, this task requires iteration over Eqs. (2)– (4). This is called *Lambda iteration* but it usually fails in optically thick regimes. This happens because photons can be trapped and scattered many times, if a single cell of the computational domain is optically thick. The ordinary Lambda iteration is not able to account for these processes on scales smaller than the spatial resolution. The idea behind the accelerated Lambda iteration (ALI), is to extract these sub-cell scattering contributions from the Lambda operator (and hence from the iteration), because we are not able to resolve them anyway. The extracted part of the ordinary Lambda operator is then put into a new *approximated* Lambda operator, which is solved quasi-analytically. Since the approximated Lambda operator usually only contains a small part of the whole Lambda operator (the subgrid part so to say), it is easy to compute, store and solve. Mathematically, the Lambda operator becomes split

$$\Lambda = (\Lambda - \Lambda^*) + \Lambda^* \tag{A.2}$$

where Λ^* denotes the approximated Lambda operator. Inserting this into Eq. A.1 and using the source function for isotropic scattering (Eq. 22), we get

$$S = \epsilon B + (1 - \epsilon)(\Lambda - \Lambda^*)S + (1 - \epsilon)\Lambda^* S \tag{A.3}$$

Since the Λ^*-operator consists of only a small part of the whole Lambda-operator, it is sparse and easy to solve. We bring it to the left-hand side:

$$[1 - (1 - \epsilon)\Lambda^*]S = \epsilon B + (1 - \epsilon)(\Lambda - \Lambda^*)S. \tag{A.4}$$

We introduce the iteration scheme, because there is still a contribution of the source function on the right-hand side. This remaining contribution can be regarded as the non-local contribution of the radiation field, which is solved by iteration. Inverting the approximated Lambda-operator then yields

$$Sn+1=[1-(1-\epsilon)\Lambda^*]^{-1}(\epsilon B+(1-\epsilon)(\Lambda-\Lambda^*)Sn). \qquad (A.5)$$

The scheme in Eq. A.5 is a combination of iteration and analytic solution. The non-local contributions (in the Lambda matrix $(\Lambda-\Lambda^*)$) are accounted for by iteration while the local subgrid scattering is handled by an inversion of the approximated Lambda operator (Λ^*). The computational cost of the inversion of the Λ^*-operator depends on its bandwidth, which determines the range on which we solve analytically. Obviously, a diagonal Λ^* is trivial to invert. But since the diagonal part of the Lambda operator describes only the local scattering in a single cell, it is not the best choice in terms of iterative performance. Usually, a tri-diagonal operator yields the best compromise between fast convergence and computational cost. But this requires the solution of a coupled set of linear equations, which is complex to implement. For now, we stay with a diagonal local Λ^*-operator, since it is the easiest one to implement and still has a tremendous effect on the convergence rate.

APPENDIX B. THE ANGULAR DISCRETIZATION USING HEALPIX

The choice of the solid angle grid is equivalent to the problem of discretizing the surface of a unit sphere. The method of characteristics requires the solution of the parameterized RTE along a large number of directions **n** depending on the anisotropy of the specific intensity $I(\mathbf{x}, \mathbf{n})$. In general, this requires a homogeneous discretization of the solid angle Ω on the 4π unit sphere. For this purpose, we use the HEALPix [5] scheme introduced by Górski et al. (2005). HEALPix ensures an optimal discretization of the unit sphere (also called *pixelation* or *tessellation*) into a number of finite solid angles $\Delta\Omega$. HEALPix in general addresses problems in which a function on domains of spherical topology has to be analyzed. The pixelation scheme was originally developed to handle large datasets generated by cosmic microwave background experiments (e.g., WMAP, Planck) and provides a software library [6] with numerous subroutines for spherical discretization and numerical analysis of functions or datasets on the sphere.

The HEALPix pixelation has a base resolution of 12 pixels in three rings around the poles and the equator of the unit sphere each covering the same area. Based on these base pixels, the resolution is refined by dividing each base pixel into N_{side}^2 subpixels, where N_{side} has to be a power of 2 ($N_{side} = 1, 2, 4, 8, \ldots$). The total number of pixels (assuming an isotropic refinement) is then $N_{pix} = 12 N_{side}^2$ (Buntemeyer et al., 2015) (Fig. B.18).

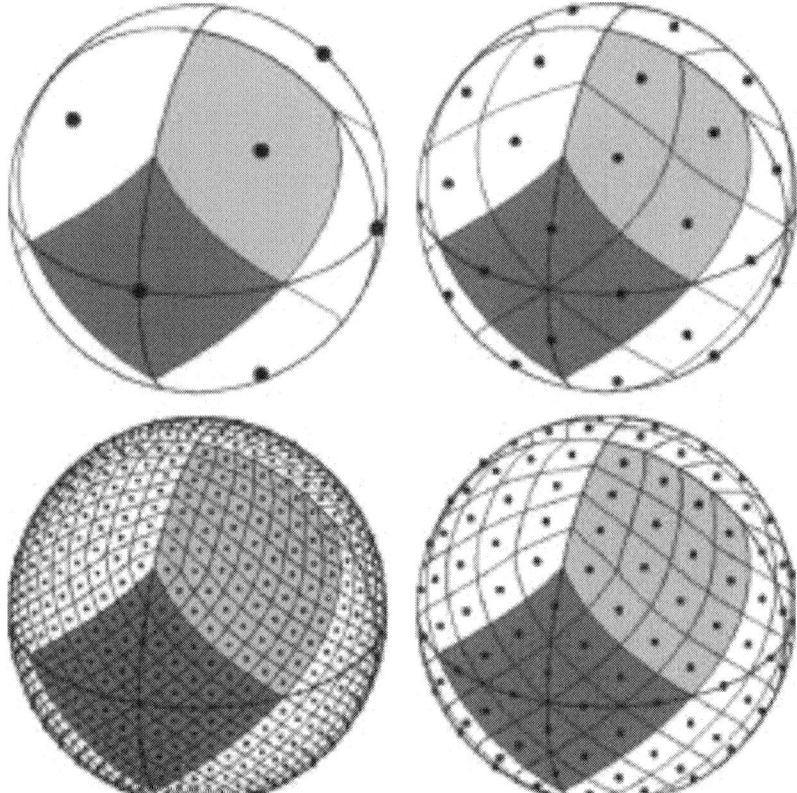

Figure B18. The HEALPix tessellation scheme, from Górski et al. (2005). The Area in light grey shows one of the eight (four north, and four south) polar base pixels and the dark grey area shows one of the four equatorial base pixels. Moving clockwise from the upper left panel the base pixels are hierarchically subdivided with the grid resolution parameter equal to $N_{side} = 1, 2, 4, 8$ and the total number of pixels $N_{pix} = 12, 48, 192, 768$.

REFERENCES

Acreman, D.M., Harries, T.J., Rundle, D.A., 2010. MNRAS 403, 1143. doi:10.1111/j.1365- 2966.2009.16199.x.

Amanatides, J., Woo, A., 1987. A fast voxel traversal algorithm for ray tracing. In: Maréchal, G. (Ed.), Proceedings of Eurographics 87. Elsevier Science Publishers, pp. 3–10

Bate, M.R., 2009. MNRAS 392, 1363. doi:10.1111/j.1365-2966.2008.14165.x.

Bate, M.R., Tricco, T.S., Price, D.J., 2013. MNRAS doi:10.1093/mnras/stt1865.

Bjorkman, J.E., Wood, K., 2001. ApJ 554, 615. doi:10.1086/321336.

Bryan, G.L., Norman, M.L., O'Shea, B.W., Abel, T., Wise, J.H., Turk, M.J., Reynolds, D.R., Collins, D.C., Wang, P., Skillman, S.W., Smith, B., Harkness, R.P., Bordner, J., Kim, J.- h., Kuhlen, M., Xu, H., Goldbaum, N., Hummels, C., Kritsuk, A.G., Tasker, E., Skory, S., Simpson, C.M., Hahn, O., Oishi, J.S., So, G.C., Zhao, F., Cen, R., Li, Y., The Enzo Collaboration, 2014. ApJS 211, 19. doi:10.1088/0067-0049/211/2/19.

Buntemeyer, L., Banerjee, R., Peters, T., Klassen, M., Pudritz, R.E., 2015. Radiation Hydrodynamics using Characteristics on Adaptive Decomposed Domains for Massively Parallel Star Formation Simulations. arXiv: 1501.04501

Chiang, E.I., Goldreich, P., 1997. ApJ 490, 368. doi:10.1086/304869.

Commerçon, B., Teyssier, R., Audit, E., Hennebelle, P., Chabrier, G., 2011. A&A 529, A35+. doi:10.1051/0004-6361/201015880

Dale, J.E., Ercolano, B., Bonnell, I.A., 2013. MNRAS 430, 234. doi:10.1093/mnras/sts592.

Davis, S.W., Stone, J.M., Jiang, Y.-F., 2012. ApJS 199, 9. doi:10.1088/0067-0049/199/1/9.

Dullemond, C.P., 2002. A&A 395, 853. doi:10.1051/0004-6361:20021300.

Dullemond, C. P., 2012. RADMC-3D: A Multi-Purpose Radiative Transfer Tool. Astrophysics Source Code Library. x1202.015

Dullemond, C.P., Monnier, J.D., 2010. ARA&A 48, 205. doi:10.1146/annurev-astro- 081309-130932.

Federrath, C., Banerjee, R., Clark, P.C., Klessen, R.S., 2010. ApJ 713, 269. doi:10.1088/0004-637X/713/1/269

Flock, M., Fromang, S., González, M., Commerçon, B., 2013. A&A 560, A43. doi:10.1051/0004-6361/201322451.

Forgan, D., Rice, K., 2013. MNRAS 433, 1796. doi:10.1093/mnras/stt736.

Fryxell, B., Olson, K., Ricker, P., Timmes, F.X., Zingale, M., Lamb, D.Q., MacNeice, P., Rosner, R., Truran, J.W., Tufo, H., 2000. ApJS 131, 273

González, M., Audit, E., Huynh, P., 2007. A&A 464, 429. doi:10.1051/0004-6361:20065486.

Goodman, A.A., Benson, P.J., Fuller, G.A., Myers, P.C., 1993. ApJ 406, 528. doi:10.1086/172465

Górski, K.M., Hivon, E., Banday, A.J., Wandelt, B.D., Hansen, F.K., Reinecke, M., Bartelmann, M., 2005. ApJ 622, 759. doi:10.1086/427976.

Hayek, W., Asplund, M., Carlsson, M., Trampedach, R., Collet, R., Gudiksen, B.V., Hansteen, V.H., Leenaarts, J., 2010. A&A 517, A49. doi:10.1051/0004-6361/201014210.

Heinemann, T., Dobler, W., Nordlund, A., Brandenburg, A., 2006. A&A 448, 731. ° doi:10.1051/0004-6361:20053120.

Jack, D., Hauschildt, P.H., Baron, E., 2012. A&A 546, A39. doi:10.1051/0004-6361/201118152.

Jiang, Y.-F., Stone, J.M., Davis, S.W., 2012. ApJS 199, 14. doi:10.1088/0067-0049/199/1/14.

Klassen, M., Kuiper, R., Pudritz, R.E., Peters, T., Banerjee, R., Buntemeyer, L., 2014. ApJ 797, 4. doi:10.1088/0004-637X/797/1/4

Kolmogorov, A., 1941. Akad. Nauk SSSR Dokl. 30, 301.

Krumholz, M.R., Klein, R.I., McKee, C.F., 2007. ApJ 656, 959. doi:10.1086/510664.

Kuiper, R., Klahr, H., Dullemond, C., Kley, W., Henning, T., 2010. A&A 511, A81. doi:10.1051/0004-6361/200912355

Larson, R.B., 1969. MNRAS 145, 271+.

Levermore, C.D., Pomraning, G.C., 1981. ApJ 248, 321. doi:10.1086/159157.

Lowrie, R.B., Edwards, J.D., 2008. Shock Waves 18, 129. doi:10.1007/s00193-008-0143- 0.

Lucy, L.B., 1999. A&A 344, 282.

Machida, M.N., Inutsuka, S.-i., Matsumoto, T., 2010. ApJ 724, 1006. doi:10.1088/0004- 637X/724/2/1006

Masunaga, H., Miyama, S.M., Inutsuka, S.-I., 1998. ApJ 495, 346. doi:10.1086/305281.

Mellon, R.R., Li, Z., 2008. ApJ 681, 1356. doi:10.1086/587542.

Mihalas, D., Weibel Mihalas, B., 1984. Foundations of radiation hydrodynamics. Oxford University Press, New York.

Min, M., Dullemond, C.P., Dominik, C., de Koter, A., Hovenier, J.W., 2009. A&A 497, 155. doi:10.1051/0004-6361/200811470.

Nakamoto, T., Umemura, M., Susa, H., 2001. MNRAS 321, 593. doi:10.1046/j.1365- 8711.2001.04008.x.

Offner, S.S.R., Klein, R.I., McKee, C.F., Krumholz, M.R., 2009. ApJ 703, 131. doi:10.1088/0004-637X/703/1/131.

Olson, G.L., Auer, L.H., Buchler, J.R., 1986. J. Quant. Spec. Radiat. Transf. 35, 431. doi:10.1016/0022-4073(86)90030-0

Olson, K.M., MacNeice, P., Fryxell, B., Ricker, P., Timmes, F.X., Zingale, M., 1999. Bulletin of the American Astronomical Society 31, 1430+.

Pascucci, I., Wolf, S., Steinacker, J., Dullemond, C.P., Henning, T., Niccolini, G., Woitke, P., Lopez, B., 2004. A&A 417, 793–805. doi:10.1051/0004-6361:20040017.

Peters, T., Banerjee, R., Klessen, R.S., Mac Low, M., Galván-Madrid, R., Keto, E.R., 2010. ApJ 711, 1017. doi:10.1088/0004-637X/711/2/1017.

Price, D.J., Bate, M.R., 2010. Magnetic fields and radiative feedback in the star formation process. In: Bertin, G., de Luca, F., Lodato, G., Pozzoli, R., Romé, M. (Eds.), Proceedings of American Institute of Physics Conference Series. In: vol. 1242. American Institute of Physics Conference Series, 1002.0650, pp. 205–218. doi:10.1063/1.3460126.

Razoumov, A.O., Cardall, C.Y., 2005. MNRAS 362, 1413. doi:10.1111/j.1365-2966.2005.09409.x.

Rijkhorst, E.-J., Plewa, T., Dubey, A., Mellema, G., 2006. A&A 452, 907. doi:10.1051/0004- 6361:20053401.

Seifried, D., Banerjee, R., Klessen, R.S., Duffin, D., Pudritz, R.E., 2011. MNRAS 417, 1054. doi:10.1111/j.1365-2966.2011.19320.x.

Seifried, D., Banerjee, R., Pudritz, R.E., Klessen, R.S., 2013. MNRAS 432, 3320. doi:10.1093/mnras/stt682.

Semenov, D., Henning, T., Helling, C., Ilgner, M., Sedlmayr, E., 2003. A&A 410, 611. doi:10.1051/0004-6361:20031279.

Stone, J.M., Mihalas, D., Norman, M.L., 1992. ApJS 80, 819. doi:10.1086/191682.

Tanaka, S., Yoshikawa, K., Okamoto, T., Hasegawa, K., 2014. A new ray-tracing scheme for 3D diffuse radiation transfer on highly parallel architectures. ArXiv eprints1410.0763.

Tobin, J.J., Hartmann, L., Chiang, H.-F., Wilner, D.J., Looney, L.W., Loinard, L., Calvet, N., D'Alessio, P., 2012. Nature 492, 83. doi:10.1038/nature11610.

Truelove, J.K., Klein, R.I., McKee, C.F., Holliman, J.H., Howell, L.H., Greenough, J.A., 1997. ApJ 489, L179+.

Trujillo Bueno, J., Fabiani Bendicho, P., 1995. ApJ 455, 646. doi:10.1086/176612.

van Noort, M., Hubeny, I., Lanz, T., 2002. ApJ 568, 1066. doi:10.1086/338949.

Walch, S.K., Whitworth, A.P., Bisbas, T., Wünsch, R., Hubber, D., 2012. MNRAS 427, 625. doi:10.1111/j.1365-2966.2012.21767.x.

Yorke, H.W., Bodenheimer, P., Laughlin, G., 1993. ApJ 411, 274.

Zhang, Y., Tan, J.C., McKee, C.F., 2013. ApJ 766, 86. doi:10.1088/0004-637X/766/2/86.

CITATION

Lars Buntemeyer, Robi Banerjee, Thomas Peters, Mikhail Klassen, Ralph E. Pudritz, Radiation hydrodynamics using characteristics on adaptive decomposed domains for massively parallel star formation simulations, New Astronomy, Volume 43, February 2016, Pages 49-69, ISSN 1384-1076, http://dx.doi.org/10.1016/j.newast.2015.07.002.

Chapter 10

Detection of Solar Neutron Events and Their Theoretical Approach

Xiao Xia Yu [a,b,c], Hong Lu [a], Guan Ting Chen [d], Xin Qiao Li [a], Jian Kui Shi [b], Cheng Ming Tan [c]

[a] Key Laboratory of Particle Astrophysics, Institute of High Energy Physics, Chinese Academy of Sciences, Beijing 100049, China
[b] State Key Laboratory of Space Weather, Chinese Academy of Sciences, 100190, China
[c] Key Laboratory of Solar Activity, National Astronomical Observatories, Chinese Academy of Sciences, Beijing 100012, China
[d] School of Economics and Management, Tsinghua University, Beijing 100084, China

ABSTRACT

Solar neutron events provide important opportunities to explore particle acceleration mechanisms using data from ground-based detectors and spacecrafts. Energetic neutrons carry crucial physics information of the acceleration site, such as energy spectrum, atmospheric elements of solar flare, scale height, convergence of the magnetic field and magnetohydrodynamic turbulence. Here 12 representative solar neutron events observed on the Earth, together with X and γ-ray observations from spacecrafts are presented. Theoretical approaches on solar neutrons that are carried out mainly through the Monte Carlo simulation are compared with the observation data, and the constraints of different theoretical models on the observations are to be summarized.

INTRODUCTION

Solar activities are intense explosions which accompany by producing high energy particles including solar protons, electrons and a small amount of neutrons. Neutrons are produced by collisions of ions in the solar photosphere, such as the interactions between the accelerated proton, heavy ion and the surrounding atmosphere as presented in Fig. 1. The processes include p–p reaction, α–α reaction, p–α reaction, p and α reaction with the surrounding heavy nuclei and their reverse reactions. Researches on the neutron and gamma rays can directly obtain the following information: the total number of accelerated particles, time evolution, angular distribution and their propagation in the flare atmosphere, etc. (Dorman, 2010, Hua et al., 2002 and Murphy et al., 2012).

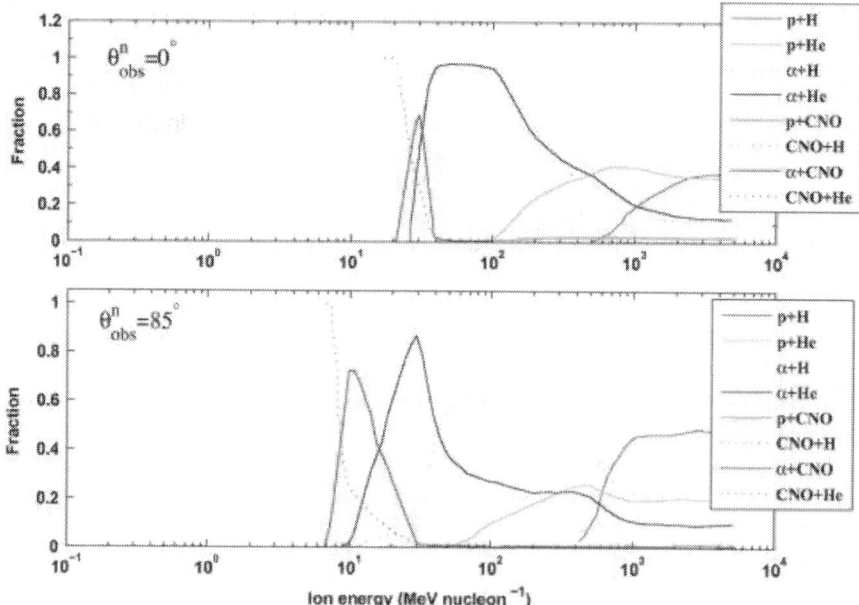

Figure 1. Eight types of neutron-producing reactions which contribute to neutron production ($E > 30$ MeV). Top panel is for a disk flare and bottom panel is for a limb flare. CNO means all the nuclear species heavier than 4He.

Theoretical researches on neutron escaping and eventually transferring to the Earth's surface were first put forward by Lingenfelter et al. (1965). Since then, Monte Carlo simulations have been developed to track a neutron until neutron was captured by hydrogen atoms in the solar atmosphere which emits 2.223 MeV neutron-capture line or escapes from the surface of the Sun (Hua and Lingenfelter, 1987a). Furthermore, they computed escaping-neutron angular distribution, energy spectrum and escaping-neutron spectrum near the Earth, considering the spiral angle of magnetic field in the solar flare loops and the magnetic mirror effect (Hua et al., 2002). The above calculations also include the relationship between the production of anisotropic neutrons, the height of the solar atmosphere, time evolution, angle and energy. Formula (1) and (2) express the neutron angular distribution and energy distribution, respectively (Hua et al., 2002).

$$\frac{d^2\sigma}{dE_n d\Omega_n} \propto exp[-(1 - \mu_n)E_n/T_0], E_n < E_{max},$$

(1)

where E_n is neutron energy, μ_n means cosine value of emitted neutron relative to the incident proton, and E_{max} is the maximal energy of neutron.

$$\frac{d\sigma}{dE_n^*} \propto (E_n^*)^\gamma exp^{-E_n^*/T_{evap}},$$

(2)

where E_n^* is neutron energy in the center of mass, T_{evap} is evaporation temperature with a value of 2.5 MeV, and $\gamma = 5/11$.

In addition, theoretical researches have also shown that time scale of neutron detected near the Earth is about 1166 s when neutron energy E_n is about 100 MeV (Hua et al., 2002). In the stationary reference frame of neutron, it takes about 1054 s to propagate from the Sun to the Earth. The average life span for the neutron decay is about 918 s, and it means that about 70 % neutron decays during the passage from the Sun to the Earth. High-energy neutrons are more likely to survive at 1 AU, and the corresponding angular distribution (its energy range is 100–1000 MeV) does not change essentially.

In observations, the detection methods of solar neutron can be divided into two types (Murphy et al., 2007): direct observation and indirect observation. Direct observations include the following cases: (i) Escaping-neutrons out of the solar atmosphere which reach the Earth can be directly captured by the Earth's orbit satellite. (ii) Ground-based neutron monitoring network can detect high-energy neutrons ($E > 200\,\text{MeV}$). In addition, indirect observations include the following cases: (i) The escaping-neutrons, which decay into protons, can be detected in the space (Evenson et al., 1983). (ii) Neutrons on the Sun, which are captured by hydrogen atoms in the photosphere, can emit 2.223 MeV rays in space, and they can be detected by the γ -ray spectrometer. Because neutrons ($E < 100\,\text{MeV}$) seriously attenuate during the propagation process of the Earth's atmosphere, they can not reach the Earth surface. Space exploration is the only method to detect neutrons ($E < 100\,\text{MeV}$), while spacecraft and ground-based detector can detect neutron ($E > 100\,\text{MeV}$) simultaneously.

THE GROUND-BASED OBSERVATIONAL CHARACTERISTICS OF SOLAR NEUTRON

Ground-level Enhancement (GLE) and Solar Neutron Events

Solar energetic particle (SEP) with energy higher than 500 MeV are identified as Ground-level Enhancements Events (GLEs) (Andriopoulou et al., 2011). There are two types of the ground-level enhancements of solar cosmic ray intensity, i.e. proton ones caused by accelerated charged particles and unusual enhancements from solar neutron (Belov and Asipenka, 2009). They are both rapid rise of short duration in the counting rates of ground-based neutron monitors. Neutron-dominated enhancements are always connected with the observable flares which distribute uniformly on the Sun disk, while the solar origin of proton-dominated enhancement often concentrates at the western heliolongitudes. Their other differences are listed in Table 1. So far 71 Ground-level Enhancements Events (GLEs) have been identified, and the most recent event recorded on 17 May 2012 (GLE71) were analyzed separately (Papaioannou et al., 2014 and Plainaki et al., 2014).

Table 1. Differences between the proton enhancement and neutron-dominated enhancement of solar cosmic ray.

Type	Solar origin	Freque ncy	Amplit ude	Duration	Anisotr opy	Energy spectrum
Proton enhancements	Protons and nuclei	Rarely	Up to 270%	Many hours	100%	Considerably changes in time
Neutron enhancements	Solar neutron	Seldo m	<25%	Some minutes	Not always	More stable

Because neutrons easily attenuate in the Earth's atmosphere, probability of capturing neutron is very low. Ground-level Enhancement (GLE) associated with solar activities which meet the following conditions can be confirmed as a solar neutron event. (i) We can eliminate that the enhancements are produced by high-energy ions, if the counts which are recorded by neutron monitors increase during the day and the other stations in the night without any enhancements. (ii) The time of counts recorded by the ground-based detectors is approaching to the corresponding emission time of hard X ray and γ ray in solar flare. (iii) The time of counts recorded by the ground-based detectors precedes the arrival time of solar proton events observed by GOES satellite, completely ruling out the interference from solar protons. (iv) The moments of counting enhancement recorded by the other solar neutron detectors (such as the solar neutron telescope and muon telescope) meet each other.

Characteristics of Solar Neutron Events Observed on the Earth

Table 2 shows the known observational features of 12 solar neutron events (1 case happened in solar cycle 21, and 4 cases occurred in solar cycle 22. The other 7 cases occurred in solar cycle 23). The observational characteristics include: (i) The soft X-ray magnitudes of solar neutron events that happened during solar cycle 21 and 22 were greater than X8 magnitude, while the soft X-ray magnitudes which occurred during solar cycle 23 can be X2.3 (2000-11-24 events) and X5.3 (2001-08-25 events), respectively. It suggests that there is no obvious correlation and threshold between the soft X-ray level of solar flare and capturing-probability of solar neutron events. There is also no correlation between the energy spectrum index of neutron and its corresponding soft X-ray level of solar flare in the solar neutron events. There is only weak correlation (about 0.46) between neutron flux on the Sun and the flux of soft X-ray in solar flare (Watanabe et al., 2005). In addition, the onset time of γ rays and the start time of solar neutron events show good correlation. The above observation results certainly do not rule out the constantly improved resolution of spacecrafts and the detection efficiency of cosmic-ray detectors during solar cycle 23, as well as the other factors such as neutron flux is larger in some solar neutron events (Shibata et al., 1993). (ii) The probability of disk flare and limb flare in triggering the solar neutron events is equal (6:6), and it is not consistent with Hua and Murphy's conclusion that the solar neutron events induced by limb flare is more likely to be observed (Hua et al., 2002, Hua and Lingenfelter, 1987a, Hua and Lingenfelter, 1987b and Murphy et al., 2007). (iii) As far as the altitude is concerned, ground-based stations which capture the solar neutron events mostly locate in more than 2200 m a.s.l. (except for the 2003-10-28 event, the height of the Namibia ground-based NM station is only 1240 m, and its soft X-ray level of solar flare with the value of X17.2). In terms of geomagnetic cut off, there's less chance of IGY neutron monitor to capture the solar neutron events (e.g. 1982-06-03 event and 1990-05-24 event). In general, higher the geomagnetic cutoff, more easy to exclude the interference from the low-energy ions. In addition, enhancement signals recorded by cosmic-ray detectors are generally above 4 σ. (iv) The solar neutron events are often accompanied by strong electromagnetic radiations, and the correlation between production time of solar neutron and hard X rays, γ ray is more closer than that with soft X-ray (Watanabe, 2005). When people calculate the neutron energy spectrum, the production time of solar neutron is generally considered to be simultaneous with nuclear deexcitation lines (4–7 MeV). The production time of 2.223 MeV γ ray usually lags about 100 s after high-energy neutrons emission time, because neutrons need some thermalization time to be captured by hydrogen atoms.

Table 2. 12 identified solar neutron events.

No.	Onset time	Flares class and location	Solar zenith angle/(°)	Location	Lon/lat	Instrument type	Altitude/m	Cut-off rigidity/GV	Enhancement	Spacecrafts	Observed characteristics	Solar cycle	References
1	1982-06-03 11:43 UT	X8/2B S09E72	25	Jungfraujoch (Switzerland); Lomnicky stit	8/47; 20/49	IGY; BNM64	3475; 2623	4.61; 3.2	6.7σ	SMM/GRS	Bursts during 11.43–11:44 UT show photon characteristics; Impulsive emission during 11:44–11:47 UT indicates the mixture characteristic of X-ray and neutron; Delayed emission during 11.47–12.06 UT presents the interaction characteristics of high-energy neutrons and GRS scintillation; Neutron energy spectrum can be fitted by the power-law spectrum with spectral index of −2.4	21	Chupp et al. (1981), Debrunner et al. (1983), Efimov et al. (1983), Chupp et al. (1987), Kudela (1990)
2	1990-05-24 20:48 UT	X9.3/18 N36W76	29	Climax (America)	254/39	IGY	3400	2.99	9σ	Granat/PHEBUS	Moreton wave appears during 20:47.40–20:47:50 UT; The first γ rays (75–95 MeV) peaks at 20:48:12 UT; Energetic neutron on the solar surface produced at 20:48:18 UT; 2.223 MeV and discrete spectral emission of γ-ray peak at 20:48.28 UT; The second γ rays (75–95 MeV) peaks at 20:48.36 UT	22	Debrunner et al. (1993), Smart et al. (1995), Debrunner et al. (1997), Terekhov et al. (1993), Vilmas et al. (2003)
3	1991-03-22 22:44 UT	X9.4/38 S26E28	21	Haleakala (America)	204/21	NM64	3030	4.38	2–3%	Granat/PHEBUS	X-ray burst begins at 22:43 UT; Radio flux of 2.7 GHz and 8.8 GHz peak at 22:44 UT; Hα intensity peaks at 22:45 UT	22	Pyle and Simpson (1991), Terekhov et al. (1996)
4	1991-06-04 (F)03:37 UT (N)03:41 UT	X12/3B N30E70	18.9; 15	Mt.Norikura (Japan)	138/36	NM64; SNT; μT	2770	11.5	5.1σ; 4.4σ; 3.3σ	CGRO/OSSE; NoBE	CME accompanied by metric-wave burst of IV type appears at 03:40:44 UT; γ-ray emission peaks at 03:42 UT; The second maximum of radio flux of 17 GHz appears at 04:25 UT	22	Murphy et al. (2007), Struminsky et al. (1994), Muraki et al. (1992)
5	1991-06-06 00:58 UT	X12/4B N33E44	26; 44.5	Mt.Norikura (Japan); Haleakala (America)	138/36; 204/21	NM64; NM64	2770; 3030	11.5; 4.38	5.2σ; 4.3σ	CGRO/BATSE; NoRP	Two γ-emissions (1–10 MeV) appear at 01:05 UT and 01:06 UT, respectively; The peak of millimetric wave (80 GHz) appears at 01:06 UT	22	Watanabe et al. (2003a), Muraki et al. (1995)
6	2000-11-24 (F)14:51 UT (N)15:08 UT	X2.3/2B N22W07	17.4	Mt.Chacaltaya (Bolivia)	292/−16	NM64	5200	12.5	5.5σ	GOES8 (0.5–4A); Yohkoh/SXT; Yohkoh/HXT; Yohkoh/GRS	Three bumps of soft X rays appear, and protons (Ep > 100 MeV) lag behind 1 h to reach the ground; Soft X-ray loop appears and gradually brighten; Hard X rays appears at the soft X-ray footpoint of flare, when soft X-ray source arises; 2.223 MeV neutron-capture line is superimposed on the composition of bremsstrahlung, and its emission duration is longer than γ-rays emission (C: 4.438 MeV) and O: 6.129 MeV)	23	Muraki et al. (2008)
7	2001-04-15 (F)13:19 UT (N)13:51 UT	X14/5C S20/W85	!	Mt.Chacaltaya (Bolivia)	292/−16	NM64	5200	12.5	3.6σ	GOES; Yohkoh/SXT; Yohkoh/HXT; Yohkoh/GRS; Geotail	The flare reaches its maximum at 13:50 UT; Ions are accelerated at 13:45 UT with the same profile of γ rays, and particles acceleration continues until 13:51 UT; Hard X-rays starts at 13:44 UT	23	Muraki et al. (2008)
8	2001-08-25 (F)16:23 UT (N)16:32 UT	X5.3/3B S17E34*	26.5	Mt.Chacaltaya (Bolivia)	292/−16	NM64	5200	12.5	4.7σ	GOES; Yohkoh/SXT; Yohkoh/HXT; Yohkoh/GRS	Flux of soft X-ray reaches its maximum at 16:45 UT; Soft X-ray loop appears after 16:45 UT and gradually brightens; Energy spectrum of hard X rays is the hardest at 16:32 UT; Weaker 2.223 MeV neutron-capture line is superimposed on the composition of bremsstrahlung, but no obvious γ rays (4–7 MeV) emission appears	23	Watanabe et al. (2003c)
9	2003-10-28 (F)09:51 UT (N)11:04 UT	X17.2/4B S16E08	9.5	Tsumeb (Namibia)	17/19	NM64	1240	9.21	6.4σ	GOES; RHESSI; INTEGRAL	Flux of soft X-ray reaches its maximum at 11:00 UT; The acceleration of the particle characteristics appears at 11:06 UT; Spectrum line of O (6.129 MeV) peaks at 11:06:15 UT; 2.223 MeV neutron-capture line emission peaks at 11:06 UT	23	Watanabe et al. (2006a)
10	2003-11-02 (F)17:03 UT (N)17:17 UT	X8.3/2B S14W56	11.5	Mt.Chacaltaya (Bolivia)	292/−16	NM64	5200	12.5	4.7σ	GOES; RHESSI; (800–7000 keV)	Flux of soft X-ray reaches its maximum at 17:25 UT; γ rays emission (4–7 MeV) peaks at 17:17 UT, but no demodulation γ rays line are observed; 2.223 MeV neutron-capture line emission peaks at 17:40 UT	23	Watanabe et al. (2006b)
11	2003-11-04 (F)19:29 UT (N)19:45 UT	X28/3B S19W83	50.5; 40.5	Haleakala (America); Mexico City (Mexico)	204/21; 261/19	NM64; NM64	3030; 2274	4.38; 8.61	7.5σ; 5.2σ	GOES; Geotail; INTEGRAL	Flux of soft X-ray reaches saturation at 19:53 UT; Hard X-rays spike appears at 19:44 UT; Bremsstrahlung emission of energetic γrays peaks at 19:45 UT; Emission of weaker 2.223 MeV neutron-capture line peaks at 19:47 UT	23	Watanabe et al. 2006a
12	2005-09-07 (F)17:17 UT (N)17:36 UT	X17/38 S06E89	17.5	Mt.Chacaltaya (Bolivia); Sierra Negra (Mexico)	292/−16; 262/19	NM64; NM; SNT	5200; 4580	12.5; 9.53	2σ	INTEGRAL; Geotail	4.4 MeV γ-ray line peaks at 17:38 UT; 2.223 MeV neutron-capture line are observed, but it is in accordance with the limb flare	23	Watanabe et al. (2007)

Note: (i) IGY and NM64 are two types of neutron monitors. (ii) NM denotes neutron monitor, and SNT means solar neutron telescope, with μT of muon telescope. (iii) The other flares are all the limb flares except for star (* indicates the disk flare). (iv) F denotes the start time of flare, and N means the production time of solar neutron in the column of onset time. (v) The following abbreviations stand for the spacecrafts, such as SMM/GRS (Solar Maximum Mission/Gamma Ray Spectrometer), CGRO/BATSE (Compton Gamma Ray Observatory/Burst And Transient Source Experiment), NoRP (Nobeyama Radio Polarimeters), GOES (Geostationary Operational Environmental Satellites), Yohkoh/SXT (Yohkoh/Soft X-ray Telescope), Yohkoh/HXT (Yohkoh/Hard X-ray Telescope), Yohkoh/GRS (Yohkoh/Gamma Ray Spectrometer), CGRO/OSSE (Compton Gamma Ray Observatory/Oriented Scintillation Spectrometer Experiment), CGRO/COMPTEL (Compton Gamma Ray Observatory/Imaging Compton Telescope), RHESSI (Reuven Ramaty High Energy Solar Spectroscopic Imager), and Geotail (Research on structure and dynamics of the Earth's magnetotail), and INTEGRAL (International Gamma-Ray Astrophysics Laboratory), and so on.

THE DETECTION OF SOLAR NEUTRON

Neutrons which reach the Earth escaping out of solar atmosphere can be directly captured by spacecrafts near the Earth orbit, while energetic neutrons (E > 200 MeV) can be observed by the ground-based neutron monitors. Part of solar neutrons on the Sun can be captured by hydrogen atoms with releasing neutron-capture line (2.223 MeV), or can also be detected by γ ray spectrometer in space.

The Ground-Based Detectors

Global coverage of the neutron-monitor network and solar neutron telescopes with synchronous muon telescopes can continuously track the Sun. Main structure and characteristics of the above detections are shown in Table 3. There are 11 neutron monitors of which elevation is above 2000 m and they are located near the equator (±40°) among them. The number of neutron monitors of which altitude is above 3000 m and the number of which are located near the equator (±40°) is 9. The quantity of solar neutron telescopes of which altitudes are above 2000 m and the number of which are located near the equator (±50°) is 7.

Table 3. Ground-based detectors of cosmic rays (neutron energy >200 MeV) (Muraki et al., 2008, Valdes-Galicia et al., 2009 and Yu et al., 2010).

Instrument name	Scientific target	Basic structure	Detection efficiency	Disadvantages
Neutron Monitor (NM)	Detection of neutrons and protons	Reflector, generator, moderator, proportional counters	20–30%	No directional capability
Solar neutron telescope	Arrival directions and energy of neutron	Proportional counters and scintillators	About 30%; arrival direction: ± 15°	Only sensitive for limited energy and speed range
Muon telescope	Cosmic ray intensity and anisotropy	Multilayer scintillator		Not yet formed all-weather monitoring network

Table 3 shows that the ground-based cosmic-ray detectors (neutron monitor, solar neutron telescope and muon telescope) have their own inherent characteristics. To some extent, they can complement each other, and provide intersectional observational characteristics of solar neutron events.

The Space-borne Instruments

There has been no space detector launched for the solar neutron so far, but spacecrafts which detect the solar radiation can indirectly provide solar neutron data as shown inTable 4. It is suggested that the spacial characteristics of solar neutron events reflect the whole band of electromagnetic radiation and particle radiation, etc., and it is not sufficient to analyze the complex physical processes only through spacial or ground-based observation data. It should be noted that SMM/GRS cannot discriminate the photons and neutrons, and CGRO/BATSE regards the moments of γ rays emission as the starting time of particle acceleration on the solar surface. The results of RHESSI show that γ-rays imaging can provide the location of ion acceleration and neutron production.

Table 4. Spacecrafts detection of solar neutron events.

Instrument name	Scientific target	Energy range	Lifetime	References
SMM/GRS	Solar γ rays, neutrons	γ rays: 300 keV–20 MeV Neutrons: >20 MeV	1980–1989	Chupp et al. (1987) Forrest et al. (1980)
CGRO/BATSE	γ rays burst	20–600 keV	Since 1991	Watanabe (2005)
CGRO/OSSE	Flare nuclear energy spectrum	γ rays: 0.05–150 MeV Neutrons: >10 MeV	Since 1991	Murphy et al. (2007)
CGRO/COMPTEL	Solar γ rays, neutrons	0.8–30 MeV	Since 1991	Watanabe (2005)
Yohkoh/SXT	Soft X-ray imaging, Temperature distribution of corona gas and magnetic structure	0.25–4 keV	1991–2001	Tsuneta et al. (1991)
Yohkoh/HXT	Hard X-ray imaging	L: 14–24 keV M1: 24–35 keV M2: 35–57 keV H: 57–100 keV	1991–2001	Kosugi et al. (1991)
Yohkoh/GRS	Energetic electron and ion acceleration	1.4–62 MeV	1991–2001	Watanabe (2005)
RHESSI	X-ray imaging, γ-rays imaging, basic processes of particle acceleration and energy release	3 keV–20 MeV	Since 2002	Lin et al. (2002)
CORONAS-F/SONG	γ rays and X rays	X rays: 0.05–100 MeV Neutrons: >20 MeV Electron: 12–100 MeV Proton: >75 MeV	Since 2001	Watanabe (2005)
INTEGRAL/SPI	γ rays	20 keV–8 MeV	Since 2002	Watanabe (2005)

Note: i) SMM/GRS, it can not discriminate the photons and neutrons. ii) CGRO/BATSE, the moments of γ-rays emission are regarded as the starting time of particle acceleration on the solar surface. iii) RHESSI, γ-rays imaging can provide the location of ion acceleration and neutron production.

THE CALCULATION OF THE NEUTRON ENERGY SPECTRUM

For the calculation of neutron energy spectrum in solar neutron events, four types of energy spectrum should be considered (Murphy et al., 2007): (i) Neutron energy spectrum of the source area. (ii) Neutron energy spectrum on solar surface. (iii) Neutron energy spectrum at the top of the Earth's atmosphere. (iv) Neutron energy spectrum above the ground-based detector.

Neutron Energy Spectrum of the Source Area

Angular distribution and energy distribution of neutrons from the source area can be expressed by formula (1) and (2), and its energy spectrum is shown in Fig. 2. Here incident particles take the typical Bessel spectrum ($\alpha T = 0.03$) and the power-law spectrum in the form of spectrum index ($S = 3.5$), respectively (Hua et al., 2002).

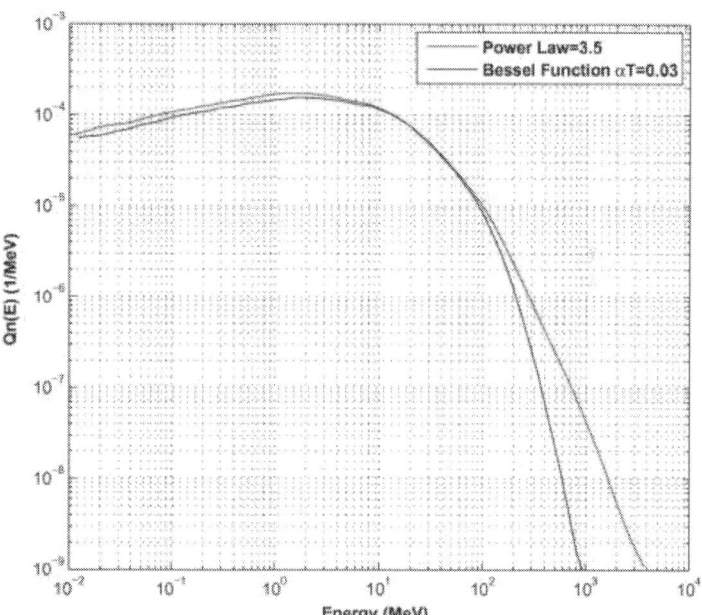

Figure 2. Neutron energy spectra are calculated by use of Hua's Monte Carlo for the incident particles having a Bessel function spectrum with $\alpha T = 0.03$ for the blue line, and a power-law spectrum with index $S = 3.5$ for the red line. (For interpretation of the references to color in this figure legend, the reader is referred to the web version of this article.)

Murphy introduced particle acceleration, magnetic-loop transport and interaction model of solar flares, and tried to explore the relationships between the acceleration parameters, physical parameters in the magnetic loops and measurements of energetic solar flare (as shown in Tables 5 and 6).

Table 5. Acceleration parameters and physical parameters in the magnetic loop model of solar flares and its values in the observation (Murphy et al., 2007 and Watanabe et al., 2009).

Parameter type	Parameter name	Loop model parameters for September 7, 2005 event
Acceleration parameters	(i) power-law spectral index, s;	(i) S = −2.0 to −6.0
	(ii) Acceleration release time history, aion(t)	(ii) 4.4 MeV line history, impulsive, etc
	(iii) Acceleration-ion composition	(iii) Impulsive, 3He/4He = 1
Physical parameters	(i) Flare loop length, L	(i) L = 36 800 km
	(ii) Level of pitch-angle scattering mean free path, λ	(ii) λ = 20 (saturated)–40,000 (none)
	(iii) Magnetic convergence index, δ	(iii) δ = 0.0 (no)–0.45 (strong)
	(iv) Flare heliocentric angle, θobs	(iv)θobs = 89.0° (S06 E89)
	(v) Atmospheric density and temperature model, n(h), T(h)	(v) Avrett model (1981)

Table 6. The corresponding relationship between the observation of high-energy flares and acceleration parameters, physical parameters in the magnetic-loop transport and interaction model (Murphy et al., 2007).

Measurement of energetic flare	Acceleration parameters	Physical parameters
Narrow deexciation line fluences and ratios	$\alpha/p, S$	Ambient abundances
Narrow deexciation line shift and shape	$\alpha/p, S$	$\delta,\lambda,\theta obs, n(h)$
Narrow deexciation line time history	$aion(t)$	$\delta,\lambda, L, n(h)$
Electron bremsstrahlung time history	$a_e(t)$	L
Neutron-capture line time history	$aion(t),\alpha/p, S$	$\delta,\lambda,\theta obs, n(h), L$
Neutron fluence at the Earth	$\alpha/p, S$	$\delta,\lambda,\theta obs, n(h)$
Neutron arrival time history at the Earth	$aion(t),\alpha/p, S$	$\delta,\lambda,\theta obs, n(h), L$

In addition, Watanabe used Hua's magnetic-loop transport and interaction model to analyze the observation data of solar neutron event on September 7, 2005, and found that this model can fit the data with appropriate values of magnetic convergence and pitch-angle scattering parameters as shown (Watanabe et al., 2009 and Hua et al., 2002) in Table 5. The heliocentric angle of solar flare (θobs) can be determined by optical image of NOAA/SEC. Furthermore, Watanabe compared the observed counting rates recorded by neutron monitors with the calculated results from Hua's transport and interaction model with $\lambda = 2000$, $\delta = 0.12$, and $s = -4.42$ (Watanabe et al., 2009), considering neutron attenuation in the Earth's atmosphere (Shibata, 1994) and detection efficiency of the neutron monitor (Clem and Dorman, 2000). It is suggested that reasonable values of physical parameters and acceleration parameters in the magnetic-loop transport and interaction models can be consistent with the observation self-consistently.

Neutron Energy Spectrum on the Solar Surface and that at the top of the Earth's Atmosphere

Flux of neutrons on the solar surface and that at the top of the Earth's atmosphere have the following relationship (Muraki et al., 1992):

$$I \odot n = \frac{L^2}{\omega(E_n)} F_{E-n},$$

(3)

where $I \odot n$ is the intensity of neutron on the solar surface in the unit of neutrons $\cdot \text{MeV}^{-1} \text{sr}^{-1}$, F_{E-n} is the flux of neutron at the top of the Earth's atmosphere in the unit of neutrons $\cdot \text{MeV}^{-1} \cdot \text{m}^2$, and $\omega(E_n) = exp(-\frac{t}{\gamma \tau})$ means the survival rate of neutron decay before reaching the Earth's surface, dependent of neutron energy. τ is the time scale of neutron decay with the value of 886 s, and $\gamma = \frac{1}{\sqrt{1-\beta^2}}$; L is one astronomy unit (1.5×10^{11} m).

Furthermore, calculation of neutron energy spectrum on the solar surface can be outlined as observation method and model method (δ model and γ model).

Observation Method

By assuming solar energetic neutrons produced with the same time profile as the γ rays, we can obtain the solar neutron energy (E_n) at that time as the following formula according to the Time of Flight Method.

$$\frac{E_n^2 + 2E_nM_n + M_n^2}{M_n^2} = \frac{c^2\Delta t^2 + L^2 + 2c\Delta tL}{c^2\Delta t^2 + 2c\Delta tL}, \qquad (4)$$

where M_n is the static mass with the value of 940 MeV, L is the distance between the Sun and the Earth, and c the light speed. According to the time profile of ground-based neutron monitor counts, we can obtain flux of solar neutron at the top of the Earth's atmosphere F_{E-n} as expressed in the following formula (Muraki et al., 1992 and Watanabe et al., 2003b).

$$F_{E-n} = \frac{\Delta N}{\Delta E_n \cdot \varepsilon(E_n) \cdot Att(E_n) \cdot S}, \qquad (5)$$

where ΔN is the excess counts of neutron monitor during this period, ΔE_n is the energy range of neutron, and $\varepsilon(E_n)$ the detection of neutron monitor dependent of neutron energy (E_n) (Clem and Dorman, 2000 and Shibata and Munakata, 2001).$Att(E_n)$ is the attenuation at the site of one neutron monitor (obtained by Monte Carlo simulation) (Debrunner et al., 1989 and Shibata, 1994), and S is the effective detecting area of one neutron monitor.

Attenuation model which solar neutrons transport in the Earth's atmosphere mainly includes Debrunner model (Debrunner et al., 1989) and Shibata model (Shibata, 1994). In terms of its accuracy, the two models have no comparability between them. Debrunner model not only includes the transportation process of solar neutrons, but also the detection efficiency of neutron monitor. But experimental results from RCNP accelerator in Japan are more inclined to support Shibata model (Koi et al., 2001). The energy spectrum of solar neutron on the solar surface (E_n^α) can be deduced from solar neutron flux (F_{E-n}) at the top of the Earth's atmosphere (Watanabe et al., 2003b) as expressed by the following formular.

$$E_{ave} = \frac{\int_{E_1}^{E_2} E_n E_n^{\alpha} dE_n}{\int_{E_1}^{E_2} E_n^{\alpha} dE_n},$$

(6)

where E_n is the kinetic energy of neutron, and α is the power-law spectral index. In addition, $Eave$ is the power-law spectral index. In addition, E_1, E_2 are the lower limit and upper limit of energy, respectively. Considering the solar neutron decay in the Earth's atmosphere and detection efficiency of neutron monitor, we get the optimal value through iterative analysis.

Model Method
Based on different assumptions about the production way of solar neutron, model method can be generally divided into two types i.e. δ model and γ model. δ model is considered that solar neutron is produced impulsively, and neutron flux on the Sun at time t can be obtained by the formula (7) (Chupp et al., 1987).

$$F \odot n(t) = L^{-2} q'(E_n, t_s) \frac{dE_n}{dt_n} \omega(E_n),$$

(7)

where $F \odot n(t)$ is the solar neutron on the Sun with the unit of neutrons $\cdot cm^{-2} \cdot s^{-1}$, and L is one astronomy unit. t_s is the flight time of neutron between the Sun and the Earth, and E_n is the kinetic energy calculated from the Time of Flight Method. $q'(E_n, t_s)$ is differential emission spectrum of neutrons on the Sun at the time of t_s, and its unit is neutrons $\cdot MeV^{-1} sr^{-1}$. $q'(E_n, t_s)$ can take the following three kinds of forms:

$$q'(E_n) \propto E_n^{-s}, E_n \leqslant E_c; q'(E_n) = 0, E_n \geqslant E_c;$$

$$q'(E_n) \propto E_n^{3/8} exp\left[-(E_n/3.26\Phi^2)^{1/4}\right];$$

$$q'(E_n) \propto exp(-E_n/E_0);$$

where $\frac{dE_n}{dt_n}$ is the energy-time dispersion relation of neutrons (Lingenfelter and Ramaty, 1967), only dependent of neutron energy, and $\omega(E_n) = exp(-\frac{t}{\gamma\tau})$ is the survival rate of solar neutrons before reaching the Earth, dependent of neutron energy (E_n).

Results compared with the space and ground observation data suggest that we can not obtain the optimum fitting results by δ function model (Chupp et al., 1987). The reasons are that (i) there are some solar neutron events of which excess count records of neutron monitors lasts for up to ten minutes. Counts of neutron monitors are significantly higher after the first photon arrived 5 min later during the impulsive emission, and it can not be explained only by the δ function model. (ii) In addition, people find that γ rays can be produced gradually through the space observation. (iii) Observations suggest that the energy spectrum of energetic neutrons is changing its intensity with time, rather than a single form. It is necessary that energetic neutrons spectrum has different forms during the pulse phase and gradual phase. We therefore take the gradual type (γ model) or hybrid type (pulse δ model and gradual γ model) for the calculation of neutron energy spectrum, and compare it with the computational results of observation method.

γ model suggests that solar neutrons are produced with the same time profile as the γ-rays nuclear deexcitation spectrum (Watanabe, 2005). If the nuclear deexcitation line is not obvious, it can be taken place with the time profile of bremsstrahlung spectrum. In fact, it is more reasonable that solar neutrons are assumed to be successively produced in some certain periods. Because γ rays emission (4–7 MeV) is the optimal indicator of solar neutron production, we should deduce the neutron spectrum on the solar surface according to the γ model. $q'(E_n, t_s)$ in the formula (7) accordingly takes the following expression (Chupp et al., 1987):

$$q'(E_n, t_s) = \sum_{t_s} q'(E_n, t_s) = K \sum_{t_s} f(t_s) q(E_n),$$

(8)

where $q'(E_n, t_s)$ is energy spectrum function of solar neutron on the Sun in the unit of neutrons $\cdot \text{MeV}^{-1} \text{ sr}^{-1}$. K is the normalization constant, and $f(t_s)$ is proportional to the time profiles of pion and γ-rays. $q(E_n)$ is the function form of solar neutron spectrum, identical to the three forms in the formula 7.

The calculation results from γ model suggest that fitting results have deficiencies, although the spacial detection (GRS) and ground-based observation from neutron monitors can limit the form of neutron energy spectrum (Chupp et al., 1987). For example, (i) We should introduce a high-energy truncation energy E_c to explain the excess counts of neutron monitors at the beginning of the event. We can reduce energy spectrum index S or increase the form of truncated spectral energy to fit the counts recorded by the neutron monitor, but neutron counts did not peak at the start time of the solar neutron events. (ii) Power-law spectrum is used to fit the data from neutron monitor, but it cannot reflect its complex structure and the time evolution of fine processes. (iii) The calculation results of hybrid model suggest that 80% of neutron emission originates from the gradual launch phase, and it is inconsistent with δ model that 100% of solar neutron is produced during the impulsive phase.

The above researches show that the calculation of neutron energy spectrum should consider neutron emission model of which energy spectrum changes with time.

Comparison of Calculation Results

According to the above two kinds of calculation methods related to solar neutron spectrum (observation method and model method), we compare solar neutron flux and spectrum index of solar neutron events in Table 7, Table 8 and Table 9 and Fig. 3. It can be inferred that observation method and model method can explain the observation parameters in the solar neutron events to some extent, but we still need to consider the influence of some other factors involved: (i) The spectrum indexes of solar neutron events which occurred during the solar cycle 23 are softer than those happened during the solar cycle 21 and 22. There are 5 solar neutron events which are observed by the neutron monitor named Mt. Chacaltaya during the solar cycle 23, and its vertical atmospheric mass is only $540\,\mathrm{g/cm^2}$ which is easy to detect weak signal with softer spectrum index. (ii) Neutron counts which reach the top of ground-based detectors in solar neutron events are almost the same, and it suggests that it is inconsistent to calculate energy spectrum of solar neutrons by using the model method. (iii) During the solar neutron events occurred on October 28, 2003, power-law spectrum index of γ model is more harder than δ model, and it conflicts with the principle of the selected model. In addition, the calculation results show that 4σ enhancement of the count rates will appear when about 10^3 neutrons inject above the detectors.

Table 7. Comparison of neutron flux obtained by the observation method in solar cycle 21 and 22 (Watanabe et al., 2005, Shibata et al., 1993, Watanabe, 2005 and Debrunner et al., 1993).

No.	Onset time	Observation method		
		Power-law spectrum index	Neutron flux on the Sun $(\times 10^{28}\ \mathrm{sr^{-1}})$	The number of neutrons at the top of the Earth's atmosphere
1	1982-06-03	−4.0 ± 0.2 (Watanabe, 2005)	690 (Watanabe, 2005)	4×10^4 (Debrunner et al., 1993)
		−3.7 ± 0.2 (Shibata et al., 1993)	8 (Debrunner et al., 1993)	
2	1990-05-24	−2.9 ± 0.1 (Watanabe, 2005)	840 (Watanabe, 2005)	4×10^5 (Debrunner et al., 1993)
		−2.6 ± 0.1 (Shibata et al., 1993)	560 (Debrunner et al., 1993)	
3	1991-03-22	−2.7 ± 0.01 (Watanabe, 2005)	11 (Watanabe, 2005)	−
		−2.4 ± 0.1 (Shibata et al., 1993)		
4	1991-06-04	−4.9 ± 1.3 (Watanabe, 2005)	73 (Watanabe, 2005)	10^4 (Debrunner et al., 1993)
		−4.9 ± 1.7 (Shibata et al., 1993)		
5	1991-06-06	$-4.6^{+0.4}_{-0.3}$ (Watanabe, 2005)	140 ± 20 (Watanabe, 2005)	−
			210 ± 20 (Watanabe, 2005)	

Table 8. Comparison of neutron flux obtained by the observation method in solar cycle 23 (Watanabe et al., 2005,Watanabe, 2005, Valdes-Galicia et al., 2009 and Watanabe et al., 2007).

No.	Onset time	Energy range of neutron (MeV)	Observation method				
			Power-law spectrum index	Neutron flux on the Sun $(\times 10^{28}\ sr^{-1})$	The number of neutrons at the top of the Earth's atmosphere (particles·m^2)	Above the detector (particles · m^2)	
6	2000-11-24	56–700	-4.2 ± 0.5	8.0	9.4×10^5	3.6×10^3	
7	2001-04-15	70–700	-4.0 ± 1.0	/	/	/	
8	2001-08-25	54–600	-3.1 ± 0.4	4.1	5.2×10^5	2.5×10^3	
9	2003-10-28	104–1400	-3.8 ± 0.4	12	2.0×10^6	1.2×10^3	
10	2003-11-02	51–180	-7.0 ± 1.3	28	2.7×10^6	3.5×10^3	
11	2003-11-04	59–913	-3.9 ± 0.5	240	3.0×10^7	3.5×10^3	
12	2005-09-07	>100	-3.8	0.6	/	/	

Table 9. Comparison of neutron flux obtained by the model method in solar cycle 23 (Watanabe et al., 2005 and Watanabe, 2005).

No.	Onset time	δ model		γ model	
		Power-law spectrum index	Neutron flux on the Sun $(\times 10^{28}\ sr^{-1})$	Power-law spectrum index	Neutron flux on the Sun $(\times 10^{28}\ sr^{-1})$
6	2000-11-24	-4.3 ± 0.2	16^{+1}_{-1}	$-4.2^{+0.3}_{-0.2}$	15^{+1}_{-2}
7	2001-08-25	$-3.0^{+0.2}_{-0.4}$	$5.8^{+0.8}_{-0.6}$	-3.1 ± 0.3	$6.6^{+0.9}_{-0.8}$
9	2003-10-28	$-4.0^{+0.4}_{-0.2}$	130^{+10}_{-10}	-2.9 ± 0.2	37^{+3}_{-4}
10	2003-11-02	$-7.9^{+0.8}_{-1.0}$	33^{+5}_{-5}	$-6.1^{+0.6}_{-0.8}$	24^{+5}_{-3}
11	2003-11-04	$-3.9^{+0.1}_{-0.2}$	550^{+40}_{-30}	$-3.6^{+0.1}_{-0.2}$	420^{+30}_{-30}

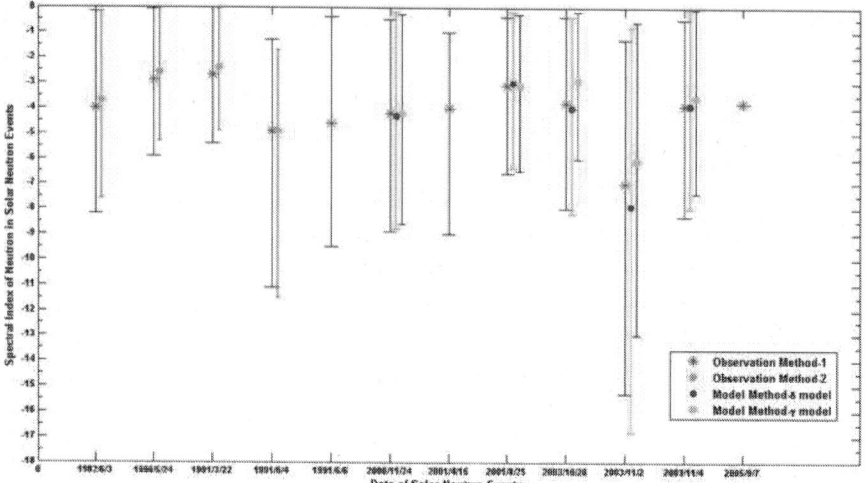

Figure 3. The simulated spectral indexes of neutrons in solar neutron events detected by ground-based neutron monitors.

Energetic neutrons, protons and γ rays can be produced in the solar neutron events. During the analysis of solar neutron event on May 24, 1990, researchers divided the whole process into the impulsive stage which neutrons and γ ray produces, emission stage which neutrons and energetic γ rays produced, and the stage that solar protons come into the near-Earth interplanetary space (Debrunner et al., 1997). They think that solar protons in the interplanetary space do not originate from the particle swarm of neutrons and γ rays, and they are the products of successive acceleration in the interplanetary space.

In general, solar neutron events often accompany by solar proton events, not vice versa. Escaping-neutron spectrum produced by ions could calculate the accelerated ion spectrum of flare (Hua et al., 2002 and Hua and Lingenfelter, 1987a) and proton number ($E > 30 \text{ MeV}$), and both of spectral indexes have the following relationship (Watanabe, 2005):

$$\alpha_n = (0.89 \pm 0.17) \times \alpha_p + (0.44 \pm 0.47), \tag{9}$$

where α_n is spectrum index of escaping-neutron, α_p is spectrum index of the accelerated ion. Here we assume that accelerated ions take the form of the power-law spectrum due to shock acceleration. We can infer from formula 9 that spectrum indexes of ion spectrum are softer than neutron energy spectrum, as shown in Table 10. We can infer that spectrum index of solar neutron is generally between 3 and 4 (except for November 2, 2003 events), and the difference between the corresponding proton spectrum and the neutron spectrum index is about −1 (e.g. spectrum index of proton is softer than that of neutron). According to the calculation, proton numbers ($E > 30\,\text{MeV}$) are about 100–1000 times of neutron flux on the solar surface (seeTable 7, Table 8 and Table 9). Because proton energy spectrum is deduced from energy spectrum of neutron, we need to consider some factors, such as energy range of neutrons, spectrum index and ion abundance in the solar atmosphere and so on. The above calculation results to some extent have certain limitations, and they need more observation samples of solar neutron events to verify furtherly.

Table 10. Comparison of neutron spectrum index, proton spectrum index and number of proton (>30 MeV) in solar neutron events (Watanabe et al., 2005, Watanabe, 2005 and Muraki et al., 2008).

No.	Onset time	Neutron spectrum index	Proton spectrum index	Number of proton (>30 MeV)/sr
1	6/3/1982	−4.0	−5.0	1033
2	5/24/1990	−2.9	−3.8	1033
3	3/22/1991	−2.7	−3.5	1031
4	6/4/1991	−4.9	−6.0	1032
5	6/6/1991	−4.6	−5.7	10^{31}–10^{32}
6	11/24/2000	−4.2	−5.2	10^{31}–10^{32}
7	4/15/2001	−3.0	−2.75	/
8	8/25/2001	−3.1	−4.0	1031
9	10/28/2003	−3.8	−4.8	10^{31}–10^{32}
10	11/2/2003	−7.0	−8.4	10^{30}–10^{31}
11	11/4/2003	−3.9	−4.9	1033

DISCUSSION AND SPECULATION

As a tool to study particle acceleration on the Sun, researches on energetic particles associated with the solar neutron events and analyses with the lag time between the peak time of energetic γ rays and enhancement time of ground-based detectors can diagnose and explore the ion acceleration mechanism, and learn the atmospheric structure and evolution of solar flare. Furthermore, we can deduce the values of the accelerated parameters and physical parameters, and study the solar-terrestrial space environment and the near-Earth space radiation by use of comprehensive utilization of multi-band and interplanetary observations. But the following problems still exist in the present researches.

I.   It is not certain whether or not solar neutrons are produced impulsively or by gradually accelerated ions, although most of solar neutron events may be fitted with an impulsive injection model. Some intensive solar flares present important evidence of extended injection, for example, April 15 2001 event, October 28 2003 event and September 7 2005 event (Valdes-Galicia et al., 2009).

II. The relation between spectrum index of solar neutron and the acceleration mechanism of flare still stays on the fit to the observation results calculated by the model. To explain the time profile of counts recorded by ground-based detectors by use of single power-law spectrum form, is difficult to present its complicated time evolution of the structure.

III. In fact, it is difficult to directly get information from neutron monitor and to determine the production time and neutron energy. Although solar neutron telescope can separate the neutral and charged particle fluxes, its ability of energy discrimination and sensitivity of speed range is limited. As a result, the probability of capturing the solar neutron events is low. The improvement on spectral and directional ability of solar neutron telescope is an important task in the future.

IV. For individual solar neutron events (such as May 24, 2009 event), we can also assume that solar neutrons are produced at the same time of electron acceleration in the solar flare, and fit the observation data (Muraki and Shibata, 1996). But X-ray observations from RHESSI satellite show that the acceleration of ions and electrons have different mechanisms (Hurford et al., 2003), it is necessary to develop the theoretical models of solar neutron emission spectrum.

V.   The detection efficiency of neutron monitors and the attenuation of solar neutron in the Earth's atmosphere are both related with neutron energy. For the calculation of neutron energy spectrum, the deviation between detection efficiency adopted from Clem (Clem and Dorman, 2000) and Shibata's measurement on the accelerator experiments (Shibata and Munakata, 2001) can be up to ±5%, and it makes demands on the calculation objectivity of neutron energy spectrum.

VI.   Neutron counts which reach the top of ground-based detectors in the solar neutron events are almost the same, and it suggests that it is inconsistent to calculate energy spectrum of solar neutron by using the model method.

ACKNOWLEDGEMENTS

The authors express their heartfelt thanks to Prof. K. Kudela for his advice to write this paper, and acknowledge the anonymous referees for the valuable suggestions which have improved the manuscript greatly. We also wish to acknowledge their enlightening and fruitful discussions, together with direct and indirect help from my teachers and colleagues, Yu Qian Ma, Huan Yu Wang, Hong Bo Hu, Feng Shi, Ping Wang, Xiao Yun Zhao, Feng Wu, Zheng Hua An and Yan Bing Xu. Systematic collections and comprehensive analyses of solar neutron events and space-borne and ground-based instruments are formidable tasks. Finally this research is supported by the Specialized Research Fund for State Key Laboratories in China and by the Fund for Key Laboratory of Solar Activity, National astronomical Observatories, Chinese Academy of Sciences (CAS).

REFERENCES

1. Andriopoulou, M., Mavromichalaki, H., Plainaki, Ch., Belov, A., Eroshenko, E., 2011. Solar Phys. 269, 155.
2. Belov, A., Asipenka, A., Dorman, L., Eroshenko, E., Kryakunova, O., Nikolayevsky, N., Shepetovz, A., Yanke V., Zhang, 2009. Proceedings of the 31st ICRC, p. 1.
3. Chupp, E.L., Forrest, D.J., Share, G.H., et al., 1983. In: Proceedings of the 18th International cosmic ray conference, vol. 10, p. 334
4. Chupp, E.L., Debrunner, H., Flückiger, E., et al., 1987. Astrophys. J. 318, 913
5. Clem, J.M., Dorman, L.I., 2000. Space Sci. Rev. 93, 335.
6. Debrunner, H., Flückiger, E., Chupp, E.L., et al., 1983. In: Proceedings of the 18th International cosmic ray conference, vol. 4, p. 75.
7. Debrunner, H., Flückiger, E.Q., Stein, P., 1989. Nucl. Instr. Meth. Phys. Res. A 278, 573
8. Debrunner, H., Lockwood, J.A., Ryan, J.M., et al., 1993. Astrophys. J. 409, 822.
9. Debrunner, H., Lockwood, J.A., Barat, C., et al., 1997. Astrophys. J. 479, 997.
10. Dorman, L., 2010. Solar Neutrons and Related Phenomena. Astrophysics and Space Science Library Springer Science Business Media B.V
11. Efimov, Y.E., Kocharov, G.E., Kudela, K., 1983. In: Proceedings of the 18th International cosmic ray conference, vol. 10, p. 276.
12. Evenson, P., Meyer, P., Pyle, K.R., 1983. Astrophys. J. 274, 875.
13. Forrest, D.J., Chupp, E.L., Ryan, J.M., 1980. Solar Phys. 65, 15.
14. Hua, X.M., Kozlovsky, B., Lingenfelter, R.E., 2002. Astrophys. J S 140, 563.
15. Hua, X.M., Lingenfelter, R.E., 1987a. Astrophys. J. 323, 779.
16. Hua, X.M., Lingenfelter, R.E., 1987b. Solar Phys. 107, 351.
17. Hurford, G.J., Schwartz, R.A., Krucker, S., et al., 2003. Astrophys. J. 595, L77.
18. Koi, T., Muraki, Y., Masuda, K., et al., 2001. Nucl. Instr. Meth. Phys. Res. A 469, 63.
19. Kosugi, T., Makishima, K., Murakami, T., et al., 1991. Solar Phys. 136, 17.
20. Kudela, K., 1990. Astrophys. J. Suppl. S. 73, 297.
21. Lin, R.P., Dennis, B.R., Hurford, G.J., et al., 2002. Solar Phys. 210, 3.
22. Lingenfelter, R.E., Flamm, E., Canfield, E.H., et al., 1965. J. Geophys. Res. 70 (4077– 4086) (4087–4095).
23. Lingenfelter, R.E., Ramaty, R., 1967. In: Shen, B.S.P. (Ed.), High-Energy Nuclear Reaction in Astrophysics A collection of articles. Benjamin, New York: Amsterdam (no. 99).
24. Muraki, Y., Murakami, K., Miyazaki, M., et al., 1992. Astrophys. J. 400, L75.
25. Muraki, Y., Sakakibara, S., Shibata, S., et al., 1995. In: Proceedings of the 24th International Cosmic Ray Conference, vol. 4, p. 175.
26. Muraki, Y., Shibata, S., 1996. In: AIP Conf Proc, vol. 374, p. 256.

27. Muraki, Y., Matsubara, Y., Masuda, S., et al., 2008. Astropart. Phys. 29, 229.
28. Murphy, R.J., Kozlovsky, B., Share, G.H., et al., 2007. Astrophys. J. Suppl. Ser. 168, 167.
29. Murphy, R.J., Kozlovsky, B., Share, G.H., et al., 2012. Astrophys. J. Suppl. Ser. 202 (3) (32p).
30. Papaioannou, A., Souvatzoglou, G., Paschalis, P., Gerontidou, M., Mavromichalaki, H., 2014. Solar Phys. 289, 423.
31. Plainaki, C., Mavromichalaki, H., Laurenza, M., Gerontidou, M., Kanellakopoulos, A., Storini, M., 2014. Astrophys. J. 785, 160 (12p).
32. Pyle, K.R., Simpson, J.A., 1991. In: Proceedings of the 22th International cosmic ray conference, vol. 3, p. 53.
33. Shibata, S., Murakami, K., Muraki, Y., 1993. In: Proceedings of the 23th International Cosmic Ray Conference, vol. 3, p. 95.
34. Shibata, S., 1994. J. Geophys. Res. 99, 6651.
35. Shibata, S., Munakata, Y., 2001. Nucl. Instr. Meth. Phys. Res. A 463, 316.
36. Smart, D.F., Shea, M.A., O'Brien, K., 1995. In: Proceedings of the 24th International cosmic ray conference, vol. 4, p. 171
37. Struminsky, A., Matsuoka, M., Takahashi, K., 1994. Astrophys. J. 429, 400.
38. Terekhov, O.V., Syunyaev, R.A., Kuznetsov, A.V., et al., 1993. Astron. Lett. 19, 65.
39. Terekhov, O.V., Sunyaev, R.A., Tkachenko, A. Yu, et al., 1996. Astron. Lett. 22, 143.
40. Tsuneta, S., Acton, L., Bruner, M., et al., 1991. Solar Phys. 136, 37.
41. Valdes-Galicia, J.F., Muraki, Y., Watanabe, K., et al., 2009. Adv. Space Res. 43, 565.
42. Vilmer, N., MacKinnon, A.L., Trottet, G., et al., 2003. Astron. Astrophys. 412, 865.
43. Watanabe, K., Muraki, Y., Murakami, K., et al., 2003a. In: Proceedings of the 28th International Cosmic Ray Conference, p. 3211.
44. Watanabe, K., Muraki, Y., Matsubara, Y., et al., 2003b. Astrophys. J. 592, 590.
45. Watanabe, K., Muraki, Y., Matsubara, Y., et al., 2003c. In: Proceedings of the 28th International Cosmic Ray Conference, p. 3179
46. Watanabe, K., 2005. Solar neutron events associated with large solar flares in solar cycle 23 (Ph.D. thesis), University of Nagoya, Nagoya.
47. Watanabe, K., Muraki, Y., Matsubara, Y., et al., 2005. In: Proceedings of the 29th International Cosmic Ray Conference, p. 101.
48. Watanabe, K., Gros, M., Stoker, P.H., et al., 2006a. Astrophys. J. 636, 1135.
49. Watanabe, K., Muraki, Y., Matsubara, Y., et al., 2006b. Adv Space Res 38, 425.
50. Watanabe, K., Sako, T., Muraki, Y., Matsubara, Y., et al., 2007. Adv. Space Res. 39, 1462
51. Watanabe, K., Lin, R.P., Krucker, S., et al., 2009. Adv. Space Res. 44, 789.
52. Yu, X.X., Lu, H., Li, G.M., Shi, F., 2010. Solar Phys. 263, 223.

CITATION

Xiao Xia Yu, Hong Lu, Guan Ting Chen, Xin Qiao Li, Jian Kui Shi, Cheng Ming Tan, Detection of solar neutron events and their theoretical approach, New Astronomy, Volume 39, August 2015, Pages 25-35, ISSN 1384-1076, http://dx.doi.org/10.1016/j.newast.2014.12.010.

Chapter 11

The Virtual Astronomical Observatory: Re-Engineering Access to Astronomical Data

R.J. Hanisch[a],[b], G.B. Berriman[c], T.J.W. Lazio [d], S. Emery Bunn[e], J. Evans f,T.A. McGlynn[g], R. Plante [h]

[a] Virtual Astronomical Observatory, 1600 14th Street NW, Suite 730, Washington, DC 20036, USA
[b] Space Telescope Science Institute, 3700 San Martin Drive, Baltimore, MD 21218, USA
[c] Infrared Processing and Analysis Center, California Institute of Technology, Pasadena, CA 91125, USA
[d] Jet Propulsion Laboratory, California Institute of Technology, Pasadena, CA 91109, USA
[e] Center for Advanced Computing Research, California Institute of Technology, Pasadena, CA 91125, USA
[f] Smithsonian Astrophysical Observatory, 60 Garden Street, Cambridge, MA 02138, USA
[g] High Energy Astrophysics Science Archive Research Center, NASA Goddard Space Flight Center, Greenbelt, MD 20771, USA
[h] National Center for Supercomputing Applications, University of Illinois, Urbana, IL 61801, USA

ABSTRACT

The US Virtual Astronomical Observatory was a software infrastructure and development project designed both to begin the establishment of an operational Virtual Observatory (VO) and to provide the US coordination with the international VO effort. The concept of the VO is to provide the means by which an astronomer is able to discover, access, and process data seamlessly, regardless of its physical location. This paper describes the origins of the VAO, including the predecessor efforts within the US National Virtual Observatory, and summarizes its main accomplishments. These accomplishments include the development of both scripting toolkits that allow scientists to incorporate VO data directly into their reduction and analysis environments and high-level science applications for data discovery, integration, analysis, and catalog cross-comparison. Working with the international community, and based on the experience from the software development, the VAO was a major contributor to international standards within the International Virtual Observatory Alliance. The VAO also demonstrated how an operational virtual observatory could be deployed, providing a robust operational environment in which VO services worldwide were routinely checked for aliveness and compliance with international standards. Finally, the VAO engaged in community outreach, developing a comprehensive web site with on-line tutorials, announcements, links to both US and internationally developed tools and services, and exhibits and hands-on training at annual meetings of the American Astronomical Society and through summer schools and community days. All digital products of the VAO Project, including software, documentation, and tutorials, are stored in a repository for community access. The enduring legacy of the VAO is an increasing expectation that new telescopes and facilities incorporate VO capabilities during the design of their data management systems.

INTRODUCTION

Beginnings

The formal Virtual Observatory (VO) program in the United States began with the 2000 Decadal Survey of the National Academy of Science, in which a National Virtual Observatory (NVO) was identified as the top priority small initiative (McKee et al., 2001).

The NVO is the committee's top-priority small initiative. NVO involves the integration of all major astronomical data archives into a digital database stored on a network of computers, the provision of advanced data exploration services for the astronomical community, and the development of data standards and tools for data mining. The committee recommends coordinated support from both NASA and the NSF, since NVO will serve both the space- and ground-based science communities.

The NVO project and parallel projects in Europe and the UK were formulated through a series of meetings, beginning with "Virtual Observatories of the Future" (Brunner et al., 2001), held at the California Institute of Technology in 2000 June.

At the 2002 conference, "Toward an International Virtual Observatory" (Quinn and Górski, 2004), held in Garching, Germany, the International Virtual Observatory Alliance[2] (IVOA) was formed with the NVO, the Astrophysical Virtual Observatory (AVO, ESO), and AstroGrid (UK) as founding partners. R. Hanisch, the then-NVO Project Manager, was the first chair of the IVOA Executive Committee. In the subsequent decade, the IVOA has grown to have 21 member national projects.

The IVOA patterned itself on the World-Wide Web Consortium[3] (W3C) and adopted its process for the development of standards (Working Drafts → Proposed Recommendations → Recommendations) with the actual standards documents developed by a set of working groups. (See Section 3.1 for more details.) A Virtual Observatory Working Group was established under Commission 5 of the International Astronomical Union (IAU) in order to give IVOA Recommendations official status within the IAU, but this process has not been used in practice since there was already global acceptance of IVOA standards.

The NVO project focused on standards and infrastructure development, working closely in the context of the IVOA, and implemented a number of prototype science applications to demonstrate the utility of the underlying VO standards. NVO also ran an active program of engagement with the astronomical community through annual summer schools of one-week duration, exhibits at American Astronomical Society meetings, and the production of a major reference book, *The National Virtual Observatory: Tools and Techniques for Astronomical Research* (Graham et al., 2007). In a demonstration of this book's value, it was translated into Mandarin by members of the VO-China project.

The NVO project was funded by the National Science Foundation's Information Technology Research program, starting in 2001, and included organizations in astronomy and computer science. Its funding came to a planned close in 2008, after demonstrating the technology framework for supporting a VO.

Program

In 2010, the successor to the NVO, the Virtual Astronomical Observatory (VAO), was begun to sustain and evolve those technologies successfully demonstrated by the NVO as part of an operating virtual observatory. While there were numerous management and logistical barriers to the establishment of the VAO, the National Science Foundation (NSF) and National Aeronautics and Space Administration (NASA) agreed to fund the project jointly, with NSF support directed through the VAO, Limited Liability Company, and NASA support provided directly to the participating NASA data centers.

The VAO, LLC, was created as a 50–50 collaboration between the Association of Universities for Research in Astronomy (AURA) and the Associated Universities, Inc. (AUI), with an independent Board of Directors. This management structure was chosen deliberately so that the VAO would be perceived as belonging to the research community and have dedicated oversight. Executive authority within the VAO was provided by the Director, who worked with a Program Manager, Project Scientist, and Project Technologist. In order to provide advice on priorities for research tools, a Science Council was established. Within the VAO, a Program Council consisting of senior management representatives from each VAO member organization was also established. The Program Council worked with the VAO management to map Science Council priorities onto available resources and expertise, and thus to develop the annual program plan. Work packages for all organizations, whether funded by NSF or NASA, were agreed with the Director and Program Manager. The program plan covered all work at all organizations regardless of the source of funding.

Table 1 shows the VAO program history and funding. As a result of two major reviews, NSF and NASA redefined program priorities and reduced the overall budget from an original plan of $27.5M ($20M NSF + $7.5M NASA) to $16.5M ($11M NSF + $5.5M NASA). In addition to simple reductions in funding, these reviews were often accompanied by recommended changes in the direction of the project, and, ultimately, the project duration was reduced by seven months. Consequently, some activities that were started or intended to be started were reduced in scope or stopped early to respond to the combination of lower funding and recommended changes in direction. A specific example of this change in direction and cessation of activities was the Time Series Search Tool (Section 2.4), which was unable to be brought to the desired level of maturity.

Table 1. VAO funding and review history.

2010 Apr	NSF Cooperative Agreement issued	$2M NSF + $1.5M NASA (FY10)
		$4M NSF + $1.5M NASA (FY11)
2010 Aug	PEP v1.0	
2010 Oct	PEP v1.1	
2011 Apr	PEP and review	$2M NSF + $1M NASA (FY12)
2012 Feb	PEP v2.0	
2012 Mar	PEP v2.1, v2.2	
2012 May	PEP v2.3	
2012 Jul	PEP and review	
2012 Sep	Decision to terminate VAO,	$2M NSF + $1M NASA (FY13)
	effective 2014 September	$1M NSF + $0.5M NASA (FY14)
	Total funding	$11M NSF + $5.5M NASA

PEP refers to the Project Execution Plan, an annual deliverable to the funding agencies. NSF's funding vehicle was a Cooperative Agreement (CA) with the VAO, LLC.

Major Accomplishments

The accomplishments of the NVO and VAO are extensive and will be described in further detail in the following sections of this paper. At a summary level, however, we note the following accomplishments:

- Major contributor to IVOA standards. Appendix B contains a list of IVOA standards to which NVO/VAO staff contributed. The list includes standards recommended by the IVOA Executive Committee and those submitted to the Executive Committee for recommendation.
- Leadership within the IVOA, within the executive, Working Groups, and Interest Groups.
- High-level science applications for data discovery, integration, analysis, and catalog cross-comparison.
- Scripting toolkits that allow scientists to incorporate VO data directly into their reduction and analysis environments.
- A robust operational environment in which VO services worldwide are routinely checked for aliveness and compliance with IVOA standards.
- Community engagement through AAS meetings, summer schools (NVO), and community days (VAO).
- Comprehensive web site with on-line tutorials, announcements, links to both US and internationally developed tools and services.
- Take up of VO standards and infrastructure within essentially every major data center and survey project in the United States, with approximately 1M VO-based data requests per month and some 2000 unique users.
- Prudent fiscal management, with overall management expenses kept below 15% and the project completed with an unspent balance of funds of less than 1% (for an $11M [lifetime] budget over 4 years).

SCIENCE APPLICATIONS

The VAO developed three science applications (Data Discovery Tool, Iris Interoperable SED Access and Analysis tool, and the Catalog Cross Comparison Service) and one prototype application (Time Series Search tool), all described in more detail below. There were also various community-led efforts that while not formal VAO projects, built upon VO standards and often involved VAO personnel in other capacities. These are also summarized below.

The motivations for developing these science applications were two-fold. First, before a standard is adopted as an IVOA Recommendation, it is expected that the Working Draft have *two* reference implementations. The objective is to ensure that the intentions of standards actually can be met in practice. In developing these science applications, the VAO provided feedback to the larger IVOA community on various aspects of IVOA standards. Second, these science applications were developed in concert with the research community, providing additional or new capabilities for addressing a variety of astronomical research questions. In the spirit that the VO is intended to enable data discovery and access for all astronomers, the applications do not serve any one observatory, wavelength, or type of user, but were intended for use by astronomers with multi-wavelength data from possibly a variety of telescopes that span the electromagnetic spectrum.

As part of a larger goal of developing an environment or "ecosystem" in which astronomical software can interact seamlessly and other tools can be contributed by the community, the development path for these science applications often included making them interoperable with other VO tools. In so doing, the VAO also provided feedback to the IVOA on the approaches toward interoperability. As a consequence of developing these applications, a number of libraries or services were developed that enable other developers to add functionality to the applications. Two examples are the SEDLIB (SED I/O library) and NED/SED service developed for Iris. Finally, by way of encouraging contributions, several collaborations (e.g., ASI Science Data Center (ASDC) archive plug-in for Iris) were fostered during VAO science applications development.

Data Discovery Tool

The Data Discovery Tool (DDT) is a web application for discovering all resources about an astrophysical object or a region of the sky (Section 3.1). Using protocols defined by the IVOA, the DDT searches those widely distributed resources that are found in the VO Registry and presents the results in a single unified Web page. In the spirit of the VAO being a working astrophysical observatory, the DDT was designed to serve as the initial steps toward a "portal", a means of discovering and accessing multi-wavelength data.

Many of the most popular US archives and catalog holdings are available for searches in the DDT, including the *Hubble Space Telescope*, *Chandra* X-ray Observatory, the Mikulski Archive for Space Telescopes (MAST), the High Energy Astrophysics Science Archive Research Center (HEASARC), Sloan Digital Sky Survey (SDSS), *Spitzer* Space Telescope, and the Two Micron All Sky Survey (2MASS), to name a few. A powerful filtering mechanism allows the user to quickly narrow the initial results to a short list of likely applicable data. Guidance on choosing appropriate data sets is provided by a variety of integrated displays, including an interactive data table, basic histogram and scatter plots, and an all-sky browser/visualizer with observation and catalog overlays (Fig. 1).

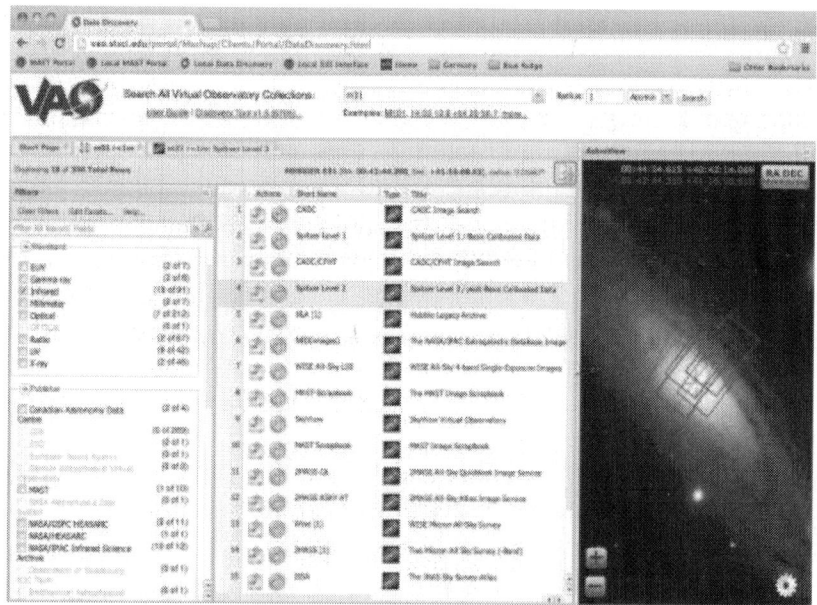

Figure.1. Appearance of the Data Discovery Tool (DDT) after a search for M31 with a radius of 1′ showing the filters (left panel) that can be applied to the search results (center panel), and the AstroView component with field-of-view overlays representing the available data sets.

The DDT was developed incrementally with the first release of the application in 2011 June. Development continued over the next two years with five incremental releases that added features and addressed any deficiencies. Web-based user documentation and training videos were developed and updated for each release.

The DDT project utilized DataScope (GSFC/NVO, McGlynn, 2007) and Astroview (STScI) and shared synergy with the MAST archive development project at STScI. IVOA standards feedback was substantial. Experience from the DDT project was used to advocate for enhanced registry metadata, table access protocol improvements, and enhanced data access protocols to ensure support for bulk queries. Staff involved in DDT development also helped to write the IVOA standard on HEALPix Multi-Order Coverage maps (Boch et al., 2014) for describing sky coverage.

Interoperable SED Access and Analysis Tool, Iris

Iris is a downloadable Graphical User Interface application that enables astronomers to build and analyze wide-band spectral energy distributions (SEDs, Doe et al., 2012,Laurino et al., 2014a and Laurino et al., 2014b). SED data may be loaded into Iris from a file on the user's local disk, from a remote URL, or directly from the NASA Extragalactic Database (NED) for analysis via the NED/SED Service. A plug-in component enables users to extend the functions of Iris. Iris utilized Sherpa (Freeman et al., 2001 and Doe et al., 2006) and Specview (Busko, 2002) as the components that performed fitting and visualization in the application. Communication between Specview and Sherpa is managed by a Simple Application Messaging Protocol (SAMP) connection (Taylor et al., 2012a and Taylor et al., 2012b). Data can also be read into Iris and can be written out via the SAMP interface (Laurino et al., 2012). A separable library for SED data input/output (SEDLib) is also included and available independently from Iris (Fig. 2).

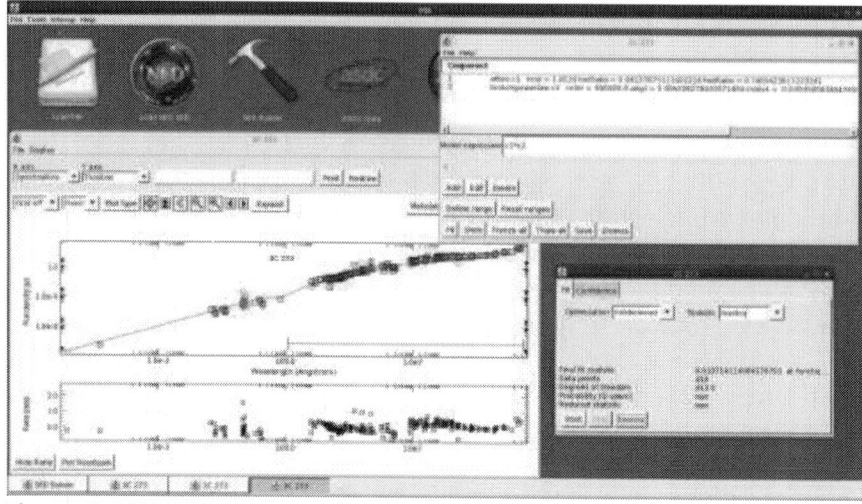

Figure. 2. VAO SED access and analysis tool Iris in operation. The Iris desktop holds the interactive windows for SED data review and analysis. Shown is a panel displaying the SED of 3C 273 with a model fit (red curve) and two panels from which the user can describe the model to fit an SED and control the fitting.

Iris was first released in 2011 October. Three incremental releases and one bug fix release followed. Iris is supported on several versions of the Mac OS X and Linux. Web-based documentation and user training videos are also provided. Iris was featured on the Astrobetter blog in 2013 September.[4]

There were two by-products of the Iris project—the NED/SED service and the SEDLib. There were collaborations with several groups including the ASI Science Data Center (ASDC) and CDS (Strasbourg). The collaborations led to Iris desktop plug-in services to access the respective SED data holdings (Laurino et al., 2013). The project provided feedback to the IVOA on the SAMP protocol, allowing for inclusion of a full SED into a single file extension, to TOPCAT (Taylor, 2005 and Taylor, 2011) for better support for SED plots, and inspired work toward a Virtual Observatory Data Model Language (VODML) by lead Iris developer O. Laurino.

Scalable Cross-Comparison Service

The Scalable Cross Comparison (SCC) Service performs fast positional cross-matches between an input table of up to 1 million sources and common astronomical source catalogs for a user-specified match radius. The service returns a list of cross-identifications to the user. The output is a composite table consisting of records from the first table, joined to all the matching records in the second table, and the angular distance and position angles of the matches (Fig. 3).

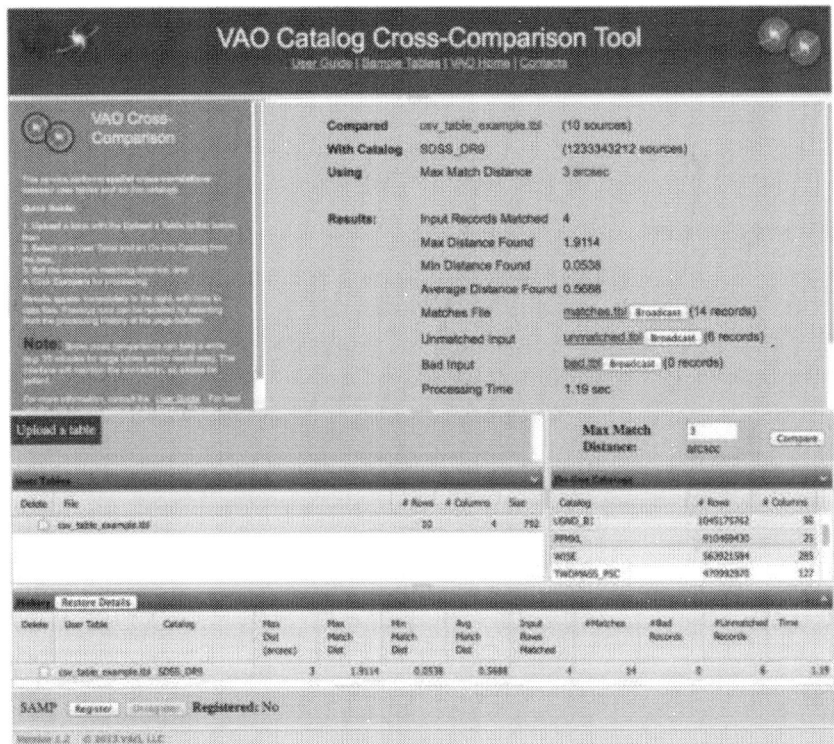

Figure 3. The Scalable Cross Comparison Service. Shown are results for a user-uploaded table cross-matched with the Sloan Digital Sky Survey Data Release 9 catalog.

The first release of the Scalable Cross Comparison Service was in 2012 January and was supported with three upgrades over the next 1.5 years. The indexing schemes that support large catalog cross-matching were provided by the Infrared Processing and Analysis Center (IPAC) and later adapted to the Wide-field Infrared Survey Explorer (WISE) and *Spitzer* projects.

Time Series Search Tool

The Time Series Search Tool finds and retrieves time series data from three major archives and analyzes them with the NASA Exoplanet Archive periodogram application. The application was a prototype developed to demonstrate that the IVOA standards of the time-series protocol and data model met the needs of such a tool. The development of the Time Series tool ended after the first VAO re-plan.

Lessons Learned

The VAO science applications group was distributed across multiple institutions, andEvans et al. (2012) described the management strategy this group. A key element of developing successful applications amongst a distributed group is managing unknowns. As the VAO science application lead might be unaware of the entire set of tasks assigned to an individual outside of the VAO efforts, coordinating task assignments and making organizational material and schedules easily available was important.

The VAO implemented a relatively lightweight process, tracked in a Wiki-based environment, in order to focus the distributed team on the requirements, design, and implementation of the applications. In addition to the developers themselves, a science stakeholder was assigned to each application and was key to bringing the view of the user to the development process. The stakeholder provided requirements, developed science use cases, handled technical questions, advised on development priorities, and performed unit tests. The use cases drove development and provided an opportunity to assess priorities and make course corrections. Internal product deliveries provided a test and assessment loop, and incremental releases (rather than one big software release) ensured that development was progressing as expected. A team lead managed priorities, schedule, and communication within the group.

Frequent communication was essential to ensuring that issues were resolved quickly and the team was working toward a common vision. The distance gap of distributed teams needs to be managed diligently. The VAO Wiki provided easy-to-access project information so that a team member could resume work quickly if he or she were sidetracked due to external project responsibilities. This process enabled the group of developers working on a project, at a distributed set of institutions and working on a part-time basis, to perform their tasks and collaborate efficiently (Evans et al., 2012).

Community Developments

During the course of the VAO, there were diverse, community-led efforts to develop VO software. (In some cases, these efforts started during the NVO era, but continued into the VAO project.) Often these involved VAO personnel, either in the role of "consultants" or who were engaged through their work on other projects.

Examples of such community-led software efforts include VOEvent, a protocol for notifications or "alerts" from and between observatories (White et al., 2006); Montage, a user-controlled tool for generating science-quality image mosaics (Berriman et al., 2003); and seleste,[5] a tool designed to provide uniform access to distributed VO databases.

STANDARDS AND INFRASTRUCTURE

The core of the VAO program was the development of software to support the IVOA standards for discovery and access to distributed data. Key components of the VAO infrastructure include the resource registry (the collection of metadata describing on-line data collections and services), the data access layer protocols (images, spectra, tables, databases) and their validation tools, a distributed authentication service ("single sign-on"), and applications programming interfaces either built-in to existing software packages or available stand-alone that allow researchers to develop their own VO-enabled scripts. Much of the VAO infrastructure is now incorporated into the data services of major data centers using VAO-provided software libraries.

The VAO Infrastructure in Context

Fig. 4 shows the VO architecture (Arviset and Gaudet, 2010). In this diagram, the VO infrastructure serves as a bridge between data providers and users, and that bridge is supported by standards. On the provider side, data is connected into the infrastructure through standard services that present that data in terms of standard data models. On the other side, users are connected to the infrastructure via generic tools that understand the VO standards. Tools are no longer tied to a single archive, but rather can talk to any and all archives that speak the common VO language.

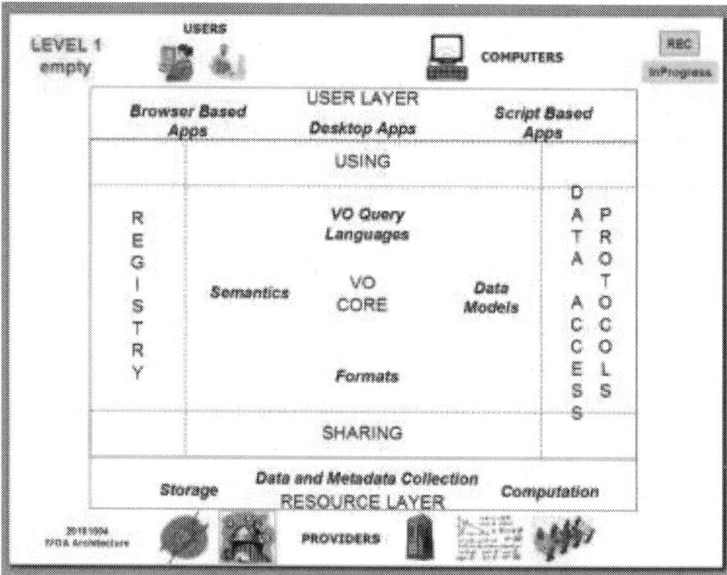

Figure 4. Virtual Observatory architecture (Arviset and Gaudet, 2010). Users appear at the top of the figure and data providers and computational resources are at the bottom, connected by the VO bridge. The VO bridge itself comprises the registry of data providers and data services, the data access protocols for discovering and retrieving data, and the core infrastructure of query languages, data models, data formats, and semantic definitions.

Providing the ability to discover and access data of interest is a significant motivation for the structure of the VO architecture. Fig. 5 illustrates the discovery framework. Registries represent the first step for data discovery in this framework. A registry is a database containing descriptions of data collections and services available in the VO (Demleitner et al., 2014). Conceptually a VO registry is similar to a "name server" for domain name service (DNS) on the Internet (Mockapetris, 1987).

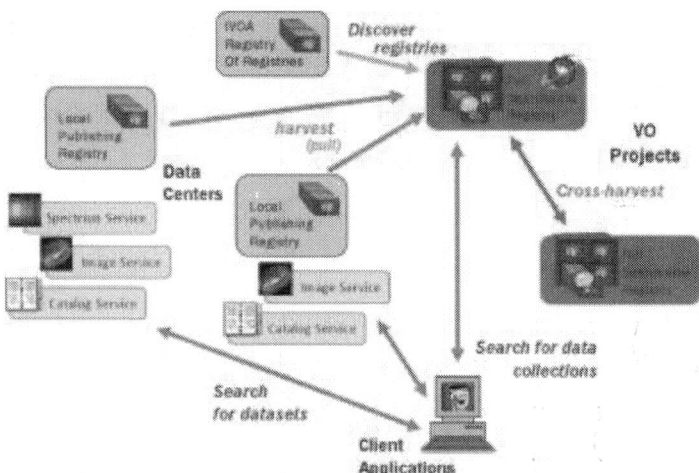

Figure. 5. Data discovery in the Virtual Observatory.

There is no single master or central registry; however, there are registries called full searchable registries that aim to have descriptions of all the data collections, archives, and service providers known to the VO from around the world. This type of registry can populate itself through a process known as harvesting; it starts by contacting a special "boot-strapping" registry (run by the VAO) called the Registry of Registries that will return to it all of the other known registries in the VO ecosystem. In order for a new registry to enter the VO, it must be registered with the Registry of Registries.

Most of the registries within the VO are publishing registries. A registry of this type is typically run by a data center that uses it to advertise the data collections and services that it offers to the VO. The full searchable registry contacts each of the publishing registries and pulls descriptions of all the data collections and services provided by the data center. At this point the full searchable registry is populated with descriptions of all of the resources known to the VO. Periodically it will re-query the other registries to obtain any new resources or other changes since the last harvest.

With an up-to-date full searchable registry available to it, a client application (e.g., Section 2) can discover any data known to the VO. It starts by asking the registry for a list of collections and services from each of the data centers that might have data relevant to the user's science question. Most of the services will be standard data access services for finding and downloading images, spectra, or catalog information from a particular archive or collection. The application can then send a query to all of the matching services to get back lists of available data sets. By browsing the returned metadata for these data sets, the user can choose which data sets to download.

VAO and the IVOA

Much of the work the VAO conducted in advancing standards was through engagement with the IVOA. The role of the IVOA is two-fold: first, to coordinate the efforts of all of the VO projects around the world, and second, to serve as a standards body for establishing VO interoperability.

From the IVOA's beginnings, the NVO and VAO were leaders in shaping the VO's global architecture and the standards that enable it, reflecting the significant data holdings of US institutions. NVO/VAO staff members served as chairs or vice-chairs of key IVOA working groups (Appendix C). The impact of this leadership is also seen in the standard documents; most of the IVOA recommendations across all of the areas of the VO have featured NVO/VAO team members either as first authors, secondary lead authors, editors, or major contributors (Appendix B).

The VAO produced many of the key reference implementations— software that demonstrates a standard in action and proves its viability. During the NVO era, there was a vigorous international debate regarding the character of the VO Registry and whether it should be relatively "coarse-grained" or "fine-grained", in terms of the amount of detail stored in the VO Registry. (See below.) The NVO created the first implementations of registries with several different architectures. The

VAO was instrumental in demonstrating data access services through software packages like DALServer (Section 3.4.1) and TAPServer (Section 3.4.2).

The NVO/VAO led the IVOA in the development of service validators. A validator is an application that checks whether another service is compliant with VO standards. A validator performs this check by sending a series of queries to a VO service and examining the response to assess whether it follows all of the rules and recommendations described in the standard. The NVO developed the first validators in the IVOA to assist data providers, allowing them to check their data access services and fix any problems before publishing them to the VO. These NVO validators quickly became critical pieces of VO infrastructure and were continued by the VAO (Section 3.4.3); and other projects joined in to contribute validators for other service standards.

In other areas, though, the VAO benefited from international developments. Not only did the VAO benefit from technical comments, there were multiple occasions in which the VAO could produce a library or tool ultimately more rapidly because some of the initial development had been done by international partners (e.g., the development of the single sign-on capability, initially developed by Astrogrid).

The Registry

As described above, a VO registry is a database containing descriptions of data collections, archives, services, and other resources, and it represents the first step in data discovery. The NVO/VAO established itself as an early leader in the area of registries. In addition to creating some of the first registries, the VAO operated the Registry of Registries (RofR) on behalf of the IVOA. The RofR allows searchable registries to bootstrap their collection of resource descriptions.

The NVO and its IVOA partners developed several different types of registries with several different implementations. There was considerable debate over the registry design, and whether it should be "fine-grained" or "coarse-grained". A fine-grained registry contains detailed metadata about the datasets available at a VO resource (for example, it might contain the right ascension and declination of all observed positions in an archive). A coarse-grained registry would only contain information about the general sky coverage of an archive. The advantage of a fine-grained registry is that one need not query distributed resources explicitly to determine if they have data of interest, whereas with a coarse-grain registry data discovery is a two-step process. The problem with a fine-grained registry, however, is that many data collections are dynamic, so that any metadata cache has to be updated continuously. Also, the structure of a fine-grained registry will necessarily be much more complicated, and harvesting of metadata between fine-grained registries could easily become inefficient. Despite the efficiencies for search and discovery offered by fine-grained registries, the VO currently operates with coarse-grained registries. The VAO consolidated support around the coarse-grained, full searchable registry service at the Space Telescope Science Institute. There are ongoing efforts to build a fine-grained registry for mostly static data collections.

VAO Directory Service

As part of the VAO's production registry, a Web-browser-based front end called the Directory Service[6] was provided. This tool is particularly useful for discovering collections and services related to a topic. By entering keywords into the search input box, the tool will return a list of resources whose description contains those keywords (Fig. 6).

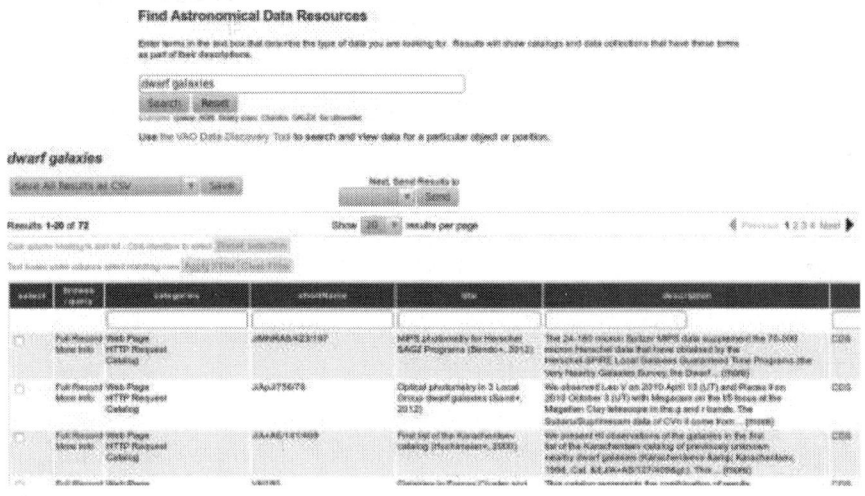

Figure 6. Results from a search query submitted through the VAO Directory Service. The directory service results allow one to browse the descriptions, filter results, download matching descriptions in VOTable format, and when the resource is a standard data access service, even send it a sky position-based query.

Registry Upgrades

Over the last two years of the VAO project, an updating program was conducted to overhaul the underlying registry database and update it to support the latest IVOA registry metadata standards. This overhaul was also necessary to support a new standard for searching registries. This new standard leverages an existing IVOA standard for querying complex databases called the Table Access Protocol (TAP), for which client software already exists. (The TAP standard did not exist when the first registry search interfaces were standardized.) Completing this upgrade was critical to maintaining the registry in the eventual post-VAO era.

A final effort conducted in the VAO project was to complete registry curation activities aimed at improving the descriptive content of the registry. In particular, a specific approach was implemented to registering resources intended to make registry searches more effective and their results less confusing. This approach has recently been accepted as a best practice by the IVOA Registry Working Group. The VAO curation work started with an inventory of existing resources by publisher, followed by developing a set of recommendations for improving the resource descriptions that brings them into line with the best practice. The new registry resource publishing tool (described below) will be instrumental in communicating these recommendations to the publisher.

Publishing Registries and the Resource Publishing Tool

A publishing registry is the vehicle for making a resource available to the VO. In particular, it can create new descriptions of resources and share them with the rest of the VO through the harvesting process. A data center, which may curate a number of data collections and offer a variety of services to access them, may operate their own publishing registry. Because such a registry does not need to serve end users directly, operating one is much simpler than running a searchable registry. During the NVO project, the VORegistry-in-a-Box product was developed that provides a simple but compliant publishing registry implementation through which a data center can maintain its own resource descriptions in-house. This product is still in production use within the VO (including by the Registry of Registries), and the VAO continued its support.

A searchable registry can also support the publishing function, which the VAO Registry at STScI does. In particular, it maintains resources descriptions on behalf of data providers who only have a few resources to share, relieving them from having to run their own publishing registry. In order to enable this feature, the VAO created the Resource Publishing Tool, a browser-based application that allows a data provider to create and share resource descriptions through the VAO Registry. It features a guided interface that steps a data provider through the process of describing a resource, prompting for metadata along the way. The tool also can check for the validity of values as they are entered, alerting the user of any problems. Draft descriptions can be saved for updating and publishing later, and already-published resource entries can be updated with this tool. Various techniques are used to minimize the amount of typing required to create a useful resource description. While the VAO Registry will share records created through this tool, the descriptions are considered "owned" by the user. Thus, to control access, the publishing tool uses the VAO Single Sign-On Services (described below).

Data Access

Standard services that allow users to find and access to data from an archive are part of the VO architecture known as the data access layer (DAL; Fig. 4). In the VO architecture, there is a standard service for each type of dataset; e.g., the Simple Image Access protocol (SIAP, Tody and Plante, 2009) enables discovery and downloading of images from an archive, and the Simple Spectral Access (SSAP, Tody et al., 2012) protocol enables access to spectra. In this section, we describe the four different "toolkits" or new protocols that the VOA developed for improved data access within the VO.

DALServer

In order to help data providers share their data collections through standard VO services, the VAO created the DALServer Toolkit, a Java-based software package. When first developed as part of the NVO project, it served as a platform for developing reference implementations of standard VO services (like SIA and SSA) that demonstrated features of the standards. At about the same time, both Astrogrid and ESO were developing data access toolkits, and some of this development fed into the VAO concept.

During the VAO's final year, specific efforts were made to enhance the toolkit for use directly by data providers; this effort was considered "productization", as it focused on making the toolkit easier to use. The focus was on a simple class of use cases in which a small data provider had a simple catalog or a simple collection of images or spectra that they wished to share. By just editing configuration files and running a few scripts, the provider could deploy fully compliant VO services with no programming required. For more complicated situations, such as for a data center that might already operate custom data access services through their own data management system, they could use the underlying DALServer Library application programming interface (API) to adapt the VO services to their local infrastructure.

The first production release of the DALServer provided support for the four "simple" standards for data access recommended by the IVOA: namely, Simple Cone Search (SCS, for simple position-based querying of object and observation catalogs, Williams et al., 2008), Simple Image Access Protocol (SIAP, for finding images), Simple Spectral Access Protocol (SSAP, for finding spectra), and Simple Line Access Protocol (SLAP, for finding rest frequencies for spectral line emissions, Salgado et al., 2010). Toward the end of the VAO project, DALServer was extended to operate on multidimensional data sets (Section 3.4.4).

TAPServer

The Table Access Protocol (TAP) is an IVOA standard for querying complex catalogs that may be made up of several tables (e.g., the 2MASS catalog). When a TAP service is connected to a catalog, users can create complex, SQL-like queries that can join metadata from several tables. Such queries are critical for mining very large catalogs. Not surprisingly given its power and flexibility, a TAP service is one of the more complex IVOA standards to implement. To make deploying a TAP service easier, the VAO created the TAPServer toolkit.

Like DALServer, TAPServer is configuration file driven. That is, with no programming required one can wrap the toolkit around a collection of tables in a database and deploy it as a service accessible to the VO. Because of the VAO close-out schedule, only a limited amount of development could be completed, and there was no effort toward the "productization" of TAPServer. However, the code is included in the VAO Repository and available for community use. Some post-VAO targeted deployments are planned. For example, it will be deployed at the National Center for Supercomputing Applications (NCSA) to expose the Dark Energy Survey Source Catalog. In turn, DES scientists will be able to analyze the catalog using the seleste[7] TAP client, a tool that allows users to form complex queries with little or no knowledge of SQL.

Service Validators

During the VAO project, service validators originally developed during the NVO project were continued and expanded. These validators have a web browser interface that allows a data center to enter a service access URL and test the services compliance with the appropriate standards; the result is a listing of errors, warnings, and recommendations for improving the service. These validators share a common Java-based toolkit platform called DALValidate. They also support a programmatic interface that allowed VAO Operations to automatically test VO services. (The VAO Operations team also engages other validators developed outside of the VAO.) Supported validators include those for Simple Cone Search, Simple Image Access, Publishing Registries, and VOResource records. The DALValidate software is available through the VAO Repository.

Image Cube Access

An emerging suite of telescopes is or soon will be generating multidimensional data (often termed "image cubes"). The most general data set, produced by an instrument measuring photons, would be $[I(\alpha,\delta,v,t),Q(\alpha,\delta,v,t),U(\alpha,\delta,v,t),V(\alpha,\delta,v,t)]$, where we have described the polarization properties by the Stokes parameters(I,Q,U,V) and each polarization can be a function of position on the sky (α,δ), frequency v (or equivalently wavelength λ or energy E), and time t. Radio interferometers have naturally produced such multidimensional data sets for some time, and the commissioning of the Jansky Very Large Array (JVLA) and the Atacama Large Millimeter/submillimeter Array (ALMA) is making such data sets much more common. X-ray telescopes have, for some time, been generating data that can be considered to be extremely sparse image cubes. The introduction of integral field units (IFUs) both for ground-based telescopes and eventually for the *James Webb Space Telescope* is making these data more common at visible and infrared wavelengths.

Providing discovery of and access to multidimensional data was taken up as a key project upon endorsement of the VAO Board. The VAO also helped stimulate interest in multidimensional data within the IVOA. In the IVOA, it was recognized that although SIAP could support image cubes in a limited way, it lacked some of the metadata support and data access mechanisms needed to support the cubes being produced or soon to be produced. From the VAO perspective, not only would discovery and access to multidimensional data advance a new capability in the VO, it might also to engage the radio astronomy community more in VO activities.

The VAO produced an early prototype service that demonstrated a number of the key capabilities needed in a new standard for image cube discovery and access. This demonstration was instrumental for mapping out the strategy for an SIAP Version 2 (Dowler et al., 2014). In particular, the necessary standardization was broken down into three independent components: (1) the Image Data Model defines the semantic labels used to describe image cubes; (2) these labels are used by the SIAPV2 standard to annotate image search results; and (3) the Access Data standard defines how one can request cutouts or other transformations of image cubes.

While active in the development of the standards within the IVOA, the VAO continued prototyping access to image cube data. To ensure that the standards served the needs of real providers of image cubes, we established a collaboration with the National Radio Astronomy Observatory (NRAO). Our joint goals were first to create a real functional image cube access service based on the emerging SIAV2 draft serving real data from NRAO instruments, and second, to provide a useful architectural design along with software to support active archive operations. In this collaboration, NRAO provided the VAO project with requirements and use cases. NRAO desired an image service that could simultaneously provide data to both internal and external clients. One key client is the Common Astronomy Software Applications (CASA, Jaeger,

2008 and CASA Consortium, 2011). Viewer that needs to request small visualizable parts of a larger cube. In return, the VAO project provided NRAO with general purpose software to both deliver data over the network to clients. The DALServer product (Section 3.4.1) was extended to provide server-side support for SIAV2, and VOClient (described below) was extended to support the client. In the spring of 2014, NRAO, using VAO-provided software, successfully demonstrated a service that provides access to image cube data, including image cutouts. This service allows their archive and CASA Viewer developers to test against a functional service.

Data Sharing

A common interest among astronomers is making their data available to their colleagues. Data sharing can be essential part of a project in which team members are in different institutions, or it can be for legacy reasons to enable data re-purposing (using the data for studies not originally envisioned when the data were acquired) or for ensuring replication. While there are institutional data centers, both in the US and internationally, there are also so-called "long-tail" data, the many small collections of data products that are typically associated with published papers. Such data products tend to be highly processed by individual astronomers and are not typically available from traditional observatory or project archives.

The second key project conducted during the close-out plan was more exploratory in its full scope (though it supported an important end-user application). The VAO sought to understand how these products could be published to the VO in a low-effort way; in order to enable such access, the focus was on integrating data sharing and publishing into the overall scholarly publishing process which starts even before the first draft of a paper. Two products were developed.

SciDrive

SciDrive (Mishin et al., 2014) is a Dropbox-like cloud storage application intended for use in scientific research. It was inspired by the SDSS MyDB (O'Mullane et al., 2004) and AstroGrid MySpace (Davenhall et al., 2004) developments, and it is based primarily on the OpenStack[8] software (in particular the OpenStack Swift component for object storage). It can be accessed from a web browser in which the user is presented with a view of a personal hierarchical directory space where one may save files by dragging-and-dropping file icons into the web page interface (Fig. 7). Also available is a desktop client that can (like Dropbox) monitor a local directory and automatically upload files that are moved into it. As many researchers already do with Dropbox, SciDrive can be a simple platform for sharing data within a research group; it provides a secure means to share read–write access to a collection within a restricted group or to send one-off permissions (read or read/write) to individuals. One difference from commercial storage providers is SciDrive's ability to scale to larger collections than with the typical free versions of storage.

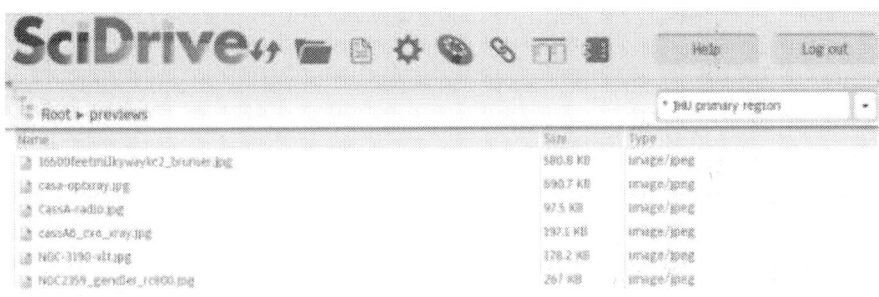

Figure. 7. The SciDrive file browser, viewed via a web browser.

SciDrive supports the VOSpace 2.0 interface, the IVOA standard for managing third party data transfers (Graham et al., 2014). This capability allows a user to seamlessly move files between different SciDrive instances (or other VO-compatible storage systems) located around the network. This feature is important for new VO capabilities in which web-based tools allow users to save application outputs to their personal space in the cloud. These outputs could be reloaded later into the tool for further analysis (e.g., as a "favorite" starting point) or loaded by other tools for synthesis with other data and analysis. A current example of this is the use of SciDrive with the Sloan Digital Sky Survey (SDSS) CasJobs: a SciDrive user can configure a directory to automatically detect uploaded table files and load them into the SkyServer database so that it can be correlated with the SDSS catalog.

The CasJobs connection highlights another unique feature of SciDrive: it supports plugins that enable special handling of certain types of data. This application is therefore a possible platform for publishing data by individual scientists and research groups. There has been experimentation with plugins that automatically extract the metadata from files that is needed to expose data to the VO. With such a feature, a research group could use SciDrive to organize a collection of data for publication. When the collection is ready for release a simple press of a button would expose the data publicly; the metadata would be automatically loaded into a database and the collection would be made available through standard IVOA services (e.g., using DALServer).

Completing this vision to a working implementation was beyond the scope of the VAO Project; nevertheless, operations and development at Johns Hopkins University (JHU) of the SciDrive platform continues (with NSF support from the Data Intensive Building Blocks program). Furthermore, VAO partners JHU and the National Center for Supercomputing Applications (NCSA) are collaborating on the emerging, community-driven initiative called the National Data Services (NDS) Consortium, which aims to address data publishing across all research fields. As the publishing scenario described above is much like one being discussed in the NDS community, we expect the development of SciDrive as a publishing platform to continue beyond the VAO project.

Single Sign-on Services

In order to restrict access to the user's personal space, SciDrive uses the VAO Login Services (Plante et al., 2012) for authentication. These services were created so that VO users could have a single login to connect any VO-compatible service or portal even when they are managed by different organizations. More than the simple convenience of a single login, a federated login system allows a user to access their proprietary data from one data center using analysis tools from another data center. A participating organization can choose to support VAO logins either as its primary identity or as an augmentation of its local authentication system. Inspired by initial developments by Astrogrid for the single sign-on capability in an astronomical context, the VAO federated login is built on the OpenID standard[9] that is in broad use across the Internet. Associated with it are all the usual services that help users manage a login: the ability to reset forgotten passwords, edit the user profile, etc. The VAO Login service also leverages an OpenID feature for sharing user information with a portal in a privacy-conscious way; this can make registering users with a portal faster and simpler. One less common feature that is important for VO applications is the ability for transparently delivering X.509 certificates to the portal. This allows a portal to access private data at another site on the user's behalf. While the service requires the user's permission to do this, it is worth noting that the user never handles the certificates directly.

The development of the VAO Login service resulted in two release software products. First, VAOSSO provides the user identity server that powers the VAO services. This software can be configured either to run as a mirror of the VAO service (for high availability) or as a completely independent service. Second, VAOLogin is a toolkit that helps portal developers add support for VAO Logins.

Current applications using the VAO Login Services include SciDrive, the VAO Registry's Resource Publishing Tool, and the VAO Notification Service. The National Optical Astronomy Observatory (NOAO) Data Archive, which currently supports the predecessor NVO Login Service, is migrating to use of the VAO Login Services to augment their own local authentication system.

Virtual Astronomy on the Desktop

A key initiative of the VAO Standards and Infrastructure program was to make VO capabilities more available from a user's local machine. Not only was the goal to make VO capabilities integrated into both new and existing desktop applications, the VAO Project sought to deliver that power directly to scientists through custom scripts that they can create to conduct their research.

Because of its growing popularity as a scripting language for scientific research, Python[10] was a major focus of our scripting support, following upon the example set by AstroGrid's python package. Further, we enabled all VO-enhanced applications and scripts running on the desktop to work together using the Simple Application Messaging Protocol (SAMP,

Taylor et al., 2012a), the IVOA standard that allows desktop and Web applications to exchange data.

VO-Enhanced Image Reduction and Analysis Facility (IRAF)

The first VAO product supporting VO on the desktop was a VO-enhanced version of IRAF (Tody, 1986 and Tody, 1993; National Optical Astronomy Observatories, 1999) developed by M. Fitzpatrick (NOAO). This included some general IRAF infrastructure enhancements including the ability to load data from arbitrary URLs as well as support for loading data in VOTable format. With these two capabilities, a suite of tasks was added to take advantage of VO services; these included an object name resolver, the ability to search the registry to find archives and services, the ability to search individual archives or catalogs, and the ability to download discovered data products. SAMP support was also added so that IRAF could send data to other non-IRAF tools running on the desktop; for example, images could be sent to Aladin (Bonnarel et al., 2000) and catalogs to TOPCAT for visualization.

VOClient

This downloadable product provides direct access to VO services outside of a Web browser. The first VOClient release featured a suite of command-line tools that enables interactive use from UNIX/Linux shell; they can also be used to create customized shell scripts. The capabilities provided by these tools include discovering archives and catalogs via the VAO registry, searching individual archives for images and spectra, downloading discovered data across multiple archives, searching catalogs by position, resolving object names to sky positions, and sending data to other desktop tools (via SAMP).

The second release of the VOClient package focused more on the underlying set of core C libraries. These libraries can be used directly to add VO capabilities to C and C++ applications (as was done for the NRAO CASA Viewer). These libraries are intended to be the basis for bindings to other languages, such as Python and Perl.[11] The Python bindings in particular were a focus of the second release (which featured a common API with PyVO, described below). Finally, the second release featured a task framework that enables easy integration of legacy software, making it callable from Python.

PyVO

This downloadable product represented a parallel effort to support Python with a slightly different focus. Through our community engagement, we found that many Python users prefer to use a pure Python implementation of a VO library, which PyVO provides, as opposed to a mixture of Python and Unix system commands. As for VOClient the audience is two-fold, the first being developers who want to integrate VO capabilities into their own Python applications. As an example, Fig. 8 shows the Ginga image browser, developed for the Subaru Telescope, to preview observatory images (Jeschke et al., 2013). Downloading of images and catalogs was an additional functionality added to the Ginga image browser using the PyVO python module.

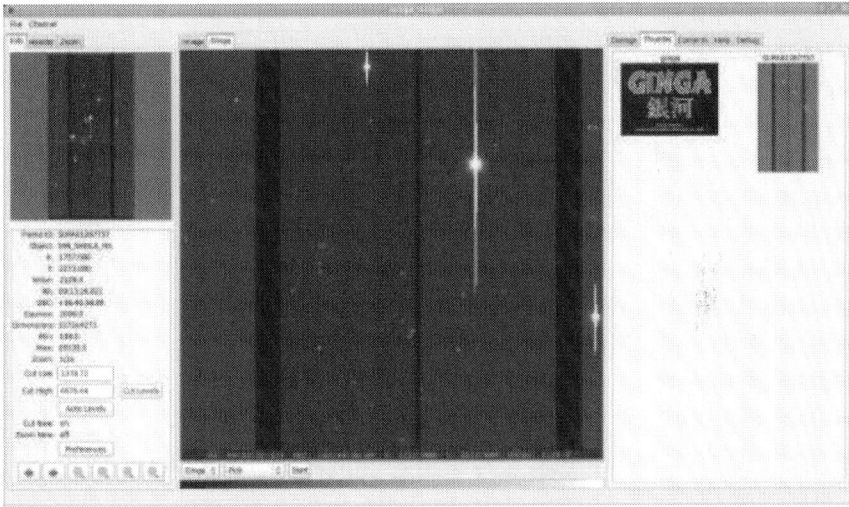

Figure. 8. Ginga Image Browser from the Subaru Telescope showing a VO plugin powered by PyVO. The rightmost panel represents a plugin that allows users to download images and catalogs from the VO for display and overlay in the viewer.

PyVO was also aimed at the growing community of research astronomers using Python to create custom scripts to carry out their research and analysis. In fact, PyVO is built on top of the widely used Astropy package (Astropy Collaboration et al., 2013), an integrated set of astronomically-oriented modules. This allows users to discover and download data and process and analyze it with the robust capabilities of Astropy. This combination is an important key to doing VO science at a large scale, as it becomes very easy to apply common processing to a vast array of data either from a single survey or from distributed collection. It also becomes possible to continuously monitor the evolving holdings of an archive or the VO in general as new data sets are added.

The first evaluation version of PyVO was released in 2013. As this release date was close to the end of the VAO Project, we wanted to ensure further use and development of PyVO beyond the Project's end. Accordingly, we explicitly employed a strategy to build a community around the PyVO package. First, GitHub[12] was used to provide a web-based code repository for future community contributions. This approach has enabled important contributions from users outside of the VAO Project; as of this writing, there are 22 issue submissions from seven external users and seven code submissions from four external users. The other part of the strategy was to establish a strong tie to the Astropy community, which is quite large and active. (In fact, this tie is responsible for much of the external participation via GitHub.) To this end, we applied for and were given status as an Astropy "affiliate package". This connection also allows PyVO to become a proving ground for migrating addition VO capabilities into Astropy.

OPERATIONS

The VAO operations effort addressed two primary goals. The first was to enable science use of the VO, in the sense of being an "operational observatory", with a focus on the VAO-developed interfaces but not exclusively. Tools must work, should work consistently, and when problems arise they must be swiftly resolved. The second goal was to enable the services needed internally for the activities of the VAO itself. VAO personnel needed reliable access to the tools needed for software design and access, user support, testing, configuration management, bug tracking, and so forth.

The VAO provided a number of science services and tools directly to the scientific community (Sections 2 and 3): its home web site, a data portal and cross-correlation tool, the Iris SED tool, downloadable VO libraries for use by clients and servers, and cloud storage and secure access protocols. Internal services included the VAO infrastructure: the JIRA ticket system, a Jenkins testing service, SVN code repository, a YouTube channel,[13] a blog, and mailing lists. The VAO also supported the IVOA Web site and document repository; these were transferred to international partners in Italy and India. The VAO software repository[14] was established to ensure that VAO-developed resources are available indefinitely.

VAO services are supported by member institutions of the VAO with significant resources hosted at each of our sites: the Smithsonian Astrophysical Observatory, JHU, MAST, HEASARC, NRAO, NOAO, Caltech, and IPAC (IRSA and NED). Most recently the software repository has used free Google cloud-based services. Elements are distributed across the country and the Internet.

Supporting such a distributed system posed (and will pose) special operational concerns. Especially for its science users, the VAO worked to ensure that elements were seen as a coherent whole: science tools need to be available at a common location, forms should have consistent look-and-feel, and everything should be clearly visible through a consistent web presence even when the web sites are on various servers.

All elements were continuously monitored and a responsible party identified for each so that issues could be rapidly and decisively addressed. The operations staff met frequently (in weekly telecons) and operational issues were rapidly escalated using an internal issue tracking software to whatever level was needed to ensure that they received the needed visibility.

Service Monitoring

All VAO services were monitored hourly and a database of all tests was continuously updated. Each service was tested to ensure not only that the service was operational, but also that it responded sensibly to some simple request. When services failed a test, they were retested 15 minutes later. If the second test also failed, a message was automatically sent to the responsible parties and to the VAO operations monitor.

A web site was available giving the current status of all operational services, and the VAO home site reflected the operations status of VAO science services so that users were immediately informed if there was an issue. Statistics were collected in and reported in biweekly periods.

Fig. 9 shows the operational status for all VAO services from spring 2011 through early summer 2014 in each bi-weekly period. The blue line shows that some of the internal VAO services–not seen directly by our science users–have had significant downtime recently. This mostly reflects in our testing and validation tools. More critically, the red line indicates only one significant lapse, in 2013 October, for the science-oriented services since early 2013. This was directly due to the shutdown of US federal services that affected NASA sites.

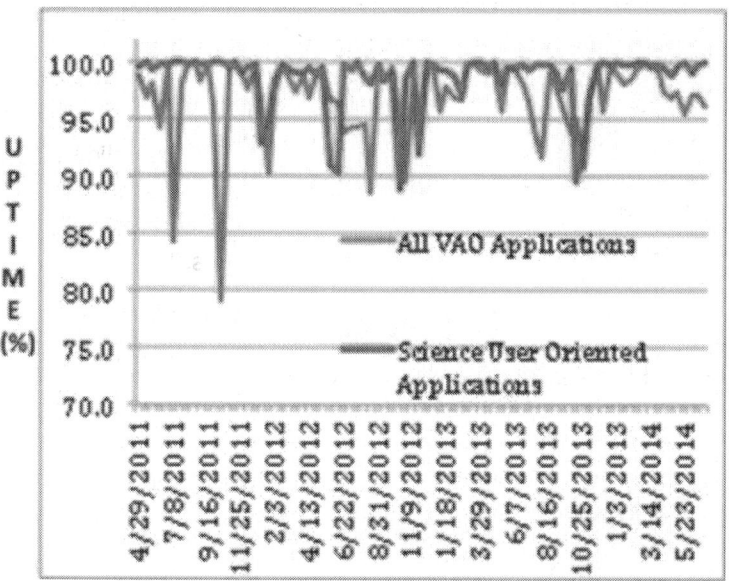

Figure. 9. Operational status for all VAO services since from spring 2011 through early summer 2014 in each biweekly period.

Monitoring and Validation of VO Data Providers

Since the effective operation of the VAO from the perspective of science users required that VAO data providers' services were available, in addition to testing the aliveness of VAO services, the VAO also monitored whether data services external to the VAO were working. Every site that published data through the VO was tested each hour. Not all published services were tested; rather a representative service from each of class of services at a site was tested. All tests were recorded and the current status of all VO sites could be seen at the VO monitoring web site. When a problem was detected, the VAO operations monitor contacted the responsible party and noted the problem. In many cases the VAO assisted such sites in rapidly bringing their services back on-line.

Occasionally a VO data-providing site is abandoned. When sites were not responsive after two months, the VAO monitoring service deprecated them in the VAO registry so that users would no longer see them in typical queries.

Each week approximately 5–10 service interruption issues were handled. In addition to testing whether services were available, the VAO also validated every published VO service using the catalog/table, image, spectral, or registry service validators. Each day approximately 300 services were validated and all validation issues were recorded in a database. This means that all published services were validated roughly once per month. Periodically, a summary report describing the VAO validation issues was prepared for each site, in order to provide concrete recommendations for resolution of validation issues.

A service that does not pass full validation can still provide valuable information, but obtaining more complete agreement with the IVOA standard ensures that tools work more robustly.

Fig. 10 shows the fraction of VO services that completely passed validation. The blue line shows all VO data providers, while the red line shows the services associated with institutions that were part of the VAO. In both cases there was a steady rise in compliance over the past several years. Two major drops in the overall compliance reflect bugs introduced at one of the major VO data providers outside the VAO. Seeing these declines our operations monitor worked with the provider, identifying specific services that were affected after initial bug fixes did not completely rectify the problem, and helped in their recovery.

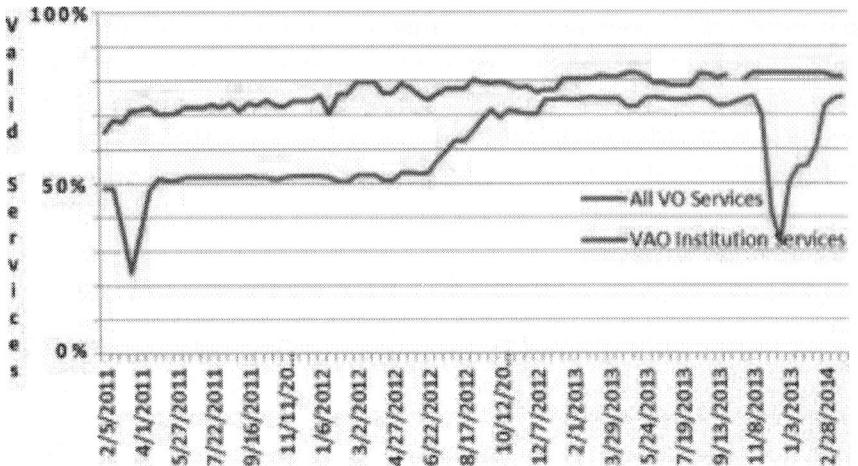

Figure. 10. Variation with time of the fraction of VO services that completely passed validation.

Post-VAO Operations

The disposition of VAO-developed services and resources is discussed in detail in other sections of this paper. Most science-oriented services will continue to be maintained by the existing institutions. The state of the internal VO services, mailing lists, documentation, blogs, and such will be maintained in the software repository. Critical infrastructure services, the web site, registry, and monitoring tools will be maintained as part of a coordinated NASA follow-on effort. This will also include at least some coordination of NASA VO operations efforts. Our experience has shown that the VO, a broadly distributed system, greatly benefits from clear and comprehensive mechanisms to identify and resolve operational issues. While the NASA follow-on effort may provide some minimal capabilities, it requires a broader national and international visibility. This is not currently something that is handled by the IVOA.

COMMUNITY ENGAGEMENT AND USER SUPPORT

During the course of the VAO, effort was undertaken to ensure that products and services delivered were robust and usable by research scientists and to reach out to the broader astronomical community. The outreach efforts aimed to expose VO products and services to potential users, to assist in the take-up of those products and services, and to gather feedback in order to assure the maximum utility of the VO for astronomical research. This section describes the full scope of the efforts.

Web Site

Fig. 11 shows the VAO web site, with an intended audience of professional astronomers and software developers. The web site was designed both to serve as an entry portal to the VAO and to provide a means for astronomers to find information about the VO—of the more than 3 million results of a search for "virtual observatory" with Google, the VAO web site is one of the top hits.

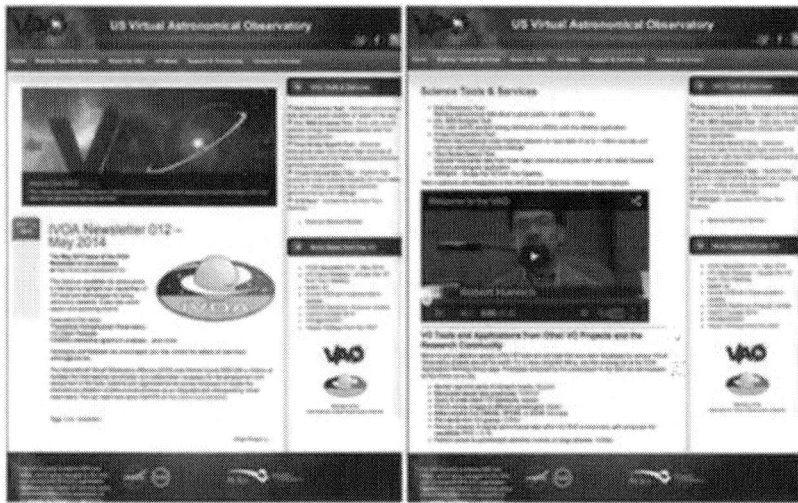

Figure 11. (*Left*) VAO web site home page. (*Right*) Science Tools and Services area within the VAO web site. Tools and services developed by the VAO appear at the top of the document, VO tools and services provided by the community appear as well.

From the perspective of the end user, the web site had two key areas. The first was "Science Tools and Services". This web document provided access to the web services or software developed by the VAO. Further, as the project began to mature, community provided tools or services began to be developed, and links to those tool or services were added.

The second area of interest for end users was "Support and Community". Analogous to the "knowledge base" that might be provided by a commercial software provider, this area was designed to help users find answers to their questions, contact other users, or submit bug reports (Fig. 12).

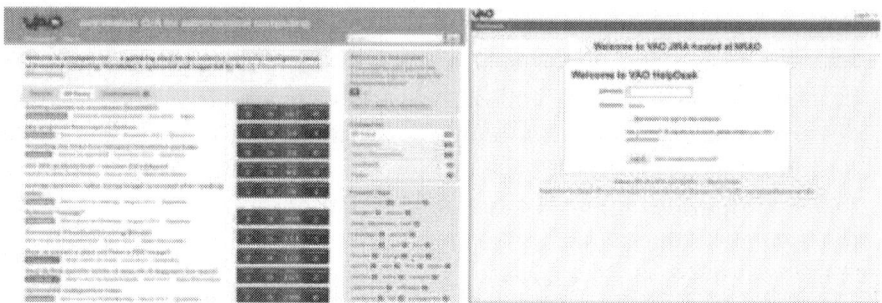

Figure 12. (Left) VAO Forum at astrobabel.com, where users could post questions and interact with other users. (Right) VAO Help Desk.

Product Testing

At the beginning of the VAO, quality control and testing activities were under the purview of User Support. The motivation for this structure was that User Support could serve as a proxy for the end user and ensure that the products and services could be used in a research setting. For most testing activities, the User Support role was to act as the coordinator of the activities and as reviewers. In addition, User Support took the lead for performing User Acceptance Testing (UAT), which was used, along with other tests and quality control reports, to prepare software release readiness reviews.

Documentation

User Support staff wrote or completed user documentation in order to help research scientists have a better understanding of VAO services and applications and how to use them. Documentation packages included deployment instructions, general descriptions, tutorials, cookbooks, and similar documents. The User Support staff and product developers also collaborated to produce video tutorials, which were then made available through a YouTube channel. All software documentation produced is available in the VAO Repository and the video tutorials remain available through YouTube.[15]

Scientific Collaborations

During the course of the project, the VAO supported the scientific or technical work of multiple individuals or collaborations. The objectives of explicitly supporting such scientific collaborations was two-fold. First, we aimed to provide examples of the VO infrastructure and capabilities being used for astronomical research. Second, the interactions with the teams were anticipated to provide feedback to the development teams for improvements to the VO infrastructure and tools. The requests for support resulted both from *ad hoc* proposals to the VAO and from a formal call for proposals that the VAO issued in 2012. The following is a summary of the projects and work supported.

- "Real-Time Analysis of Radio Continuum Images and Time Series for ASKAP" (PI: T. Murphy). This proposal requested assistance in describing multi-dimensional radio wavelength data and publishing it to the VO. Interaction with this team was used as a key use case in developing the VAO Standards and Infrastructure effort toward multi-dimensional data and in interactions with the IVOA.

- "Integration of AAVSO Data Archives into the Virtual Astronomical Observatory" (PI: M. Templeton). This proposal requested assistance in publishing data from the American Association of Variable Star Observers into the VO. The VAO provided assistance to the AAVSO, and the data are now available.

- "Cosmic Assembly/Near-infrared Deep Extragalactic Legacy Survey (CANDELS)" (PIs: S. Faber and H. Ferguson). The VAO supported the CANDELS program by distributing supernovae detections with the VOEvent network and providing access to CANDELS images through standard VO image access protocols. CANDELS supported the VAO program by providing guidance on requirements for SED building and analysis tools.

- "Brown Dwarf Candidate Identification Through Cross-Matching" (PI: S. Metchev). The VAO supported a project that continued a search for extremely red L- and T-type brown dwarfs that had begun during the NVO. It involved cross-comparing the 2MASS and SDSS catalogs to identify candidates that were followed-up with spectroscopy at the Infrared Telescope Facility, Mauna Kea. The project identified the two reddest known L dwarfs, nine probable binaries, six of which were new and eight of which likely harbor T dwarf secondary stars, and derived an estimate of the space density of T dwarfs (Geißler et al., 2011).

In addition to these scientific collaborations, a scientifically motivated sub-award was issued to produce a cross-matched multi-wavelength catalog of more than 1M objects within a 10° radius of the SMC was produced ("A Catalog of Spectral Energy Distributions of Stars in the Small Magellanic Cloud", PI: B. Madore). The catalog is in the VAO Repository, and it has been incorporated into NED with value-added content.

Booths and exhibits at American Astronomical Society meetings
American Astronomical Society (AAS) meetings, principally those occurring during the winter, are one of the focal points for the US (and international) astronomical community. During the course of the project, the VAO had exhibit booths at AAS meetings (Fig. 13). The use of an exhibit booth built on experience gained from NASA Archives and National observatories, for which it was found that substantial fractions of the community could be engaged at low cost. As an illustration of the value of an AAS meeting, people stopping at the exhibit were offered the opportunity to sign up for the VAO mailing list. At each AAS meeting, the size of the VAO mailing list increased by approximately 20%.

Figure. 13. Collage of images from the VAO Booth at the 221st American Astronomical Society Meeting, Long Beach, CA (2013 January). Also shown is one of the VAO-related posters (Kinne et al., 2013).

VAO Community Days

VAO Community Days were a series of presentations and hands-on activities designed to take the VAO to the community, demonstrate capabilities, develop and encourage new users, and obtain feedback on VO tools and services (Fig. 14). Community Days were typically structured with a morning session led by VAO team members, with the option of an afternoon session for attendees to ask more detailed questions to VAO team members or to bring in their research questions to assess how VO tools and services could assist them. Community Days were aimed initially at locations where there were a large number of astronomers with the goal of making it easy for many to attend. Table 2 lists the VO Community Days that were held. Two VAO Community Days (at the University of Washington and Cornell University) were being planned when the VAO was directed to discontinue them in preparation for its close-out activities.

Figure. 14. (Left) Example of an announcement flyer for a VO Community Day. (Right) Scene from a VAO Community Day, in which a VAO team member is demonstrating how a VO tool could be used. Both of these examples are from the Community Day held at the University of Michigan.

Table 2: VAO Community Day locations and dates.

Center for Astrophysics, Cambridge, MA	11/30/2011
Caltech, Pasadena, CA	12/9/2011
U. Arizona, Tucson, AZ	3/13/2012
U. Michigan, Ann Arbor, MI	11/14/2012
STScI, Baltimore, MD	11/27/2012

In addition to the VAO Community Days organized by the VAO, VAO Team Members also participated in similar activities organized by international organizations, including in Italy, Brazil, and Chile.

Summer Schools

During the VAO, VAO Team Members participated in summer schools organized by other institutions, often presenting lectures or developing demonstrations. The NVO project hosted four Summer Schools between 2004 and 2008. During these week-long intensive sessions, over 160 participants worked with experienced VO users and software specialists to become familiar with how to discover, access, visualize, and analyze data, and how to use the data publication and high performance computing capabilities of the VO. Those attending were introduced to VO tools and utilities and use them to accomplish a variety of research goals including data mining, multiwavelength research, and time domain astronomy. In the second half of the session small teams created their own VO-enabled data analysis applications. Students were asked to work on team-based projects using VO protocols and software in service of astronomical science. At the end of the school when the projects were presented, Summer School faculty granted awards to the five best projects. Winning projects received financial support to attend and present their work at forthcoming winter AAS meetings.

One NVO Summer School led to the production of the book *The National Virtual Observatory: Tools and Techniques for Astronomical Research* (Graham et al., 2007), which contained the lectures and tutorials from that school. The volume also included a complete set of software libraries and worked examples to guide the astronomer/software developer through the process of developing VO-enabled programs in a variety of programming languages and scripting environments. Several chapters describe research results obtained by participants in the NVO Summer Schools using VO tools and technologies.

LONG-TERM CURATION OF VAO ASSETS

The VAO Repository

The VAO is making available all its digital assets—including code, documentation, data-bases, reports—through a single Google Services repository, chosen because it is free of charge, stable, and openly accessible.[16] Its existence was announced through venues such as the AAS Newsletter, the IVOA Newsletter, and astronomy blogs and social forums. The code repository will contain all builds of the VAO software components, and all the information needed to build and use them. This content includes build instructions, release history, system requirements, license information, test results, documentation, user guides, and tutorials. The material has a common organization and look-and-feel.

Currently, the repository contains builds of the science application codes, the VAO single sign-on and login codes, and the monitoring and validation software. In addition, the repository mirrors snapshots of all software that has been committed to the VAO SVN development repository, via automated weekly up-dates.

The VAO chose not to have a software licensing policy as there will be no organization to enforce it after close-out. The software is therefore released as public domain software, while duly honoring institutional licensing policies and licensing restrictions implied by the licensing of dependent third-party software. Thus, the Iris SED builder developed at SAO is released with an Apache 2.0 license and the cross-comparison code developed at Caltech/IPAC is released with a BSD 3-clause license.

All completed documentation has been posted to the repository,[17] including software documentation, project reports, and outreach material. All project presentations and papers are also available.[18] The VAO YouTube channel, blog, Facebook page and Twitter feed will remain live.

Transition of the VAO Infrastructure to the NASA Archives

In response to a Call for Proposals issued by NASA in 2013 August, the NASA archives at STScI (MAST), IPAC (NED, IRSA, NASA Exoplanet Archives) and HEASARC submitted a proposal to sustain the core infrastructure components of the VAO within their "in-guide" budgets, beginning FY 2015 (2014 October 1). That proposal was accepted, and the NASA Archives began their activities to sustain the core VO infrastructure elements. A Project Scientist at HEASARC will coordinate VO activities between archives and report to NASA on VO-related activities.

THE VAO LEGACY

The impact of the US VO programs on the international VO can be seen in a number of ways:

Significant contributions to at least 35 IVOA standards and documents, from the first basic standards and services (VOTable, Simple Cone Search) to sophisticated data models and advanced data access protocols (Table Access Protocol, ObsCore, SIAP Version 2, ...).

Leadership of numerous IVOA Working Groups and Interest Groups, as well as leadership at the IVOA Executive level.

- A rich infrastructure for data discovery and access, with wide deployment and implementation at major data centers in the US.
- A robust operational environment in which distributed services are routinely validated against IVOA standards.
- A system of resource registries that enables discover of data and data services through the world.
- Exemplar science applications for data discovery, spectral energy distribution construction and analysis, and catalog cross-comparison.
- Desktop scripting tools including a native Python implementation.
- Cloud-based data storage for collaborative research and simple data sharing with the research community.
- Creation of a "data scientist" position at the American Astronomical Society whose responsibilities include "to help process and manage the increasing volume of digital data and to integrate it within the Virtual Observatory".
- A repository of all VAO products: software, documentation, tutorials, videos, news-letters, ...
- An increasing expectation that new telescopes and facilities incorporate VO capabilities during the design of their data management systems (e.g., Mahabal et al., 2012, Graham et al., 2012, Juric et al., 2013, Anderson et al., 2013 and Seaman et al., 2014).

However, it is more difficult to measure impact quantitatively. Since the VAO was mostly about the deployment of software tools and infrastructure services, it can be challenging to attribute data accesses to the VAO as opposed to the underlying data services. Web applications are primarily entry points to VO services; scripting environments are needed for bulk processing. In the astronomy community at least, and probably in many other disciplines, new software can take many years to penetrate the community, and even then, there is not a strong culture of software citation. For example, we find that although some 22,000 peer-reviewed papers mention the VLA radio telescope, only 68 formally acknowledge the use of AIPS and only 59 acknowledge use of CASA, the two dominant reduction and analysis packages for radio interferometry data. Remarkably (or perhaps not, given the situation for software citation) of over 13,000 peer-reviewed publications in astronomy and astrophysics published in 2013, only 4% acknowledge use of the ADS (M. Kurtz 2014, private communication) and the ADS is probably the most widely-used software system in the field. Thus, counting acknowledgments to VAO or VO tools is unlikely to reflect accurately on community take-up.

On the other hand, VAO usage logs indicate close to one million VO-based data accesses per month at US data providers, and with ~100 organizations who have published some 10,000 VO-compliant data services worldwide. VAO usage logs also show some 2000 distinct users of VAO services in the past three months (April–June 2014). The ADS lists over 2500 papers (about half of these peer-reviewed) citing "virtual observatory" in some context, and these papers are read as often and cited as often as other types of papers. Of course, without reading each and every paper one cannot be sure of the level of contamination in this sample (a paper saying "our observatory has photometry measurements of virtually thousands of stars" would count as a hit). A list of ~100 papers that make explicit use of VO tools and services are listed at http://www.usvao.org/support-community/vo-related-publications/.

The VO concept has been adopted in numerous other fields, particular in space science (with seven VxOs within NASA), plus the Virtual Solar Observatory (NASA, NSO), Planetary Science Virtual Observatory (Europe), and the Deep Carbon Virtual Observatory (Rensselaer Polytechnic Institute). The VO concept was recently endorsed by a panel of neuroscientists convened by the Kavli Foundation and General Electric as a means for improving access and interoperability to the vast data sets being collected in the European Brain Project and US Brain Initiative. VAO and IVOA participants are now playing leading roles in the international Research Data Alliance and the newly formed US National Data Services Consortium.

Lessons Learned

In looking back over the VAO project and its NVO predecessor, a number of "lessons learned" is apparent.

- Successful infrastructure is largely invisible and unappreciated. Developing metrics for measuring the success of software infrastructure is a difficult question that reaches across all scientific disciplines. The topic was, for example, discussed in detail at the 2015 NSF Software Infrastructure for Sustained Innovation (SI2) Principal Investigators meeting.[19] We urge scientists and funding agencies to investigate it collectively and develop guidelines for measuring the impact of software infrastructure. More attention should have been paid to explaining the VO infrastructure to the user community and the funding agencies, and we recommend that similar projects make such explanations a priority even in the earliest phases of development.

- Deployment of a distributed infrastructure takes considerable time. Community consensus and buy-in require early and ongoing participation. The VAO team inherited the solutions and approach of the its technology-driven predecessor, the NVO, primarily because the both projects had a common core staff. Consequently, the VAO was slow to engage the user community and deliver services that have value to astronomers in their day-to-day work. The approach eventually used, of bringing the VAO to astronomers through integration into widely used tools in consultation with the community, led to the successful delivery and take-up of the PyVO and VOClient toolkits. Nevertheless, and an earlier start would have led to more advanced and richly-featured services.

- It is important to do marketing to the research/user community, and to manage expectations. Promising too much is as bad or worse than delivering too little. Early promises for VO capabilities were overly ambitious and led to significant skepticism. This ambitious program was also the primary reason why the VAO was in the position of making a substantial number of deliveries in the final three months of the project, precisely when staff are moving on new projects at their home institutions. As a result, some deliveries were snapshots of the code rather the full featured and well documented deliveries. Thus, in addition to managing expectations, we recommend scheduling the majority of deliveries in the earlier phases of a project. Placing all the software in a central public repository ensures that all the code, whether full deliveries or snapshots, is available to the community for further development.

- The absence of a dedicated test team, led to by a dedicated test engineer, that would be available to support development and execution of test plans across the VAO increased the overhead on managing and organizing testing. This overhead arose because test teams were assembled on-the-fly from available staff, and test plans were consequently begun late in the development phase. We recommend establishing an independent test team at the start of a project, who coordinate with developers throughout the development lifecycle.
- An essential element of the VO is the Registry, within which data providers indicate what services they provide. Initially, an approach of having an easy registration process was adopted, with the consequence that some of the services registered were either of low quality or poorly maintained. It is difficult to achieve the correct balance between easy registration to encourage a substantial Registry and substantial initial quality control that results in a Registry not containing expected services.
- A distributed project has both advantages and disadvantages:

Advantage
Access to a diversity of skills and different environments for validating technical approaches and implementations.

Disadvantage
Coordination of efforts takes time; staff members have competing priorities as most were not working on VAO full-time.

For VAO the advantages outweighed the disadvantages, though there were certainly inefficiencies resulting from the distributed nature of the development work. These inefficiencies were minimized by having staff at only two or three organizations responsible for deliveries. For example, staff from SAO, STScI, and IPAC/NED developed Iris. Where appropriate, one organization was responsible for a component. HEASARC managed the operational monitoring system, for example, and NOAO managed the User Support system.

- Setting up an independent management entity such as the VAO, LLC, is a non-trivial effort, though in the VAO case it proved to be worthwhile and effective. Having a dedicated Board of Directors to provide focused advice was a great asset.
- Top-down imposition of standards is likely to fail. Attempts to turn the OpenSkyQuery protocol into an IVOA standard, for example, did not succeed because one group proposed the standard and suggested that everyone else just adopt it.
- Coordination at the international level is essential, but takes time and effort. It can be difficult to reach consensus, or even know if consensus has been reached, owing to different cultures and communications styles.
- Explicit definition of data models is important, even in cases where they seem obvious. Constructing data models after-the-fact leads to having to redefine protocols.
- Metadata collection and curation are essential and ongoing tasks, but complex, and represent a considerable investment. Across the entire VO, resources were never adequate to do a proper job of curation.

CONCLUSIONS

The NVO and VAO, working with international partners, have established the key infrastructure for data discovery, access, and interoperability in astronomy and this infrastructure extends world-wide by virtue of collaboration with the IVOA. This infrastructure is both widely adopted and heavily used, although because of the nature of infrastructure people are often unaware that they are using the VO. The IVOA has also developed a rich body of standards–45 in all–in the remarkably short period of 12 years, and the international VO efforts remain strong. Through the transfer of VAO assets to NASA, with open source software and documentation, the VAO legacy will be preserved and, we hope, enhanced. The VAO legacy will also be protected through the establishment of the US Virtual Observatory Alliance under the AAS.

ACKNOWLEDGMENTS

The VAO program would not have been possible without the financial support of the National Science Foundation (AST-0834235) and NASA (NNX13AC07G to STScI/MAST), and it was supported NASA/HEASARC. Funding at IPAC has been provided by a grant from the National Aeronautics & Space Administration (NASA) to the Jet Propulsion Laboratory, operated by the California Institute of Technology under contract to NASA. We appreciate the wise guidance of the Board of Directors of the VAO, LLC, and the VAO Science Council, and we are grateful for feedback from the astronomical community that helped us improve our science tools and infrastructure. Part of this research was carried out at the Jet Propulsion Laboratory, California Institute of Technology, under a contract with the National Aeronautics and Space Administration.

Foremost, we acknowledge the dedication, commitment, and excellence of the VAO project team. We brought together the best of the best, from nine different organizations, and through pursuit of common goals created a data management infrastructure that has brought about a sea change in how we manage and share data in astronomy and that has become a model for data management in many other disciplines. We express our gratitude D. De Young (deceased, 2011 December) for his astute guidance throughout the NVO Project and in the initial phases of the VAO.

This research has made use of NASA's Astrophysics Data System Bibliographic Services.

APPENDIX A. VAO INSTITUTIONS

The VAO was operated as a limited liability company, funded by the National Science Foundation with coordinated funding provided by the National Aeronautics and Space Administration. Table A.3 lists the institutions engaged in the scientific and technical development work of the VAO; business management was provided by the Associated Universities, Inc. (AUI).

Table A.3: Participating VAO institutions.

NSF	NASA
California Institute of Technology (Caltech)	High Energy Astrophysics Science Archive Research Center (HEASARC)
Johns Hopkins University (JHU)	Infrared Processing and Analysis Center, California Institute of Technology (IPAC)
National Center for Supercomputing Applications (NCSA)	Jet Propulsion Laboratory, California Institute of Technology (JPL)
National Optical Astronomy Observatory (NOAO)	Space Telescope Science Institute (STScI)
National Radio Astronomy Observatory (NRAO)	
Smithsonian Astrophysical Observatory (SAO)	

Institutions are listed according to which agency provided the significant funding for VAO work.

Appendix B. IVOA Standards

This appendix lists International Virtual Observatory Alliance standards and recommendations for which VAO Team Members were identified either as authors or editors. Standards and recommendations are listed in reverse chronological order of adoption.

- "VOTable Format Definition," Version 1.3, IVOA Recommendation, 20 September 2013 (F. Ochsenbein, R. Williams, C. Davenhall, M. Demleitner, D. Durand, P. Fernique, D. Giaretta, R. Hanisch, T. McGlynn, A. Szalay, M. Taylor, A. Wicenec).
- "Data Access Layer Interface," Version 1.0, IVOA Recommendation, 29 November 2013 (P. Dowler, M. Demleitner, M. Taylor, D. Tody).
- "IVOA Registry Relational Schema," Version 1.0, IVOA Proposed Recommendation, 27 February 2014 (M. Demleitner, P. Harrison, M. Molinaro, G. Greene, T. Dower, M. Perdikeas).

- "MOC—HEALPix Multi-Order Coverage map," Version 1.0, IVOA Proposed Recommendation, 10 March 2014 (T. Boch, T. Donaldson, D. Durand, P. Fernique, W. O'Mullane, M. Reinecke, M. Taylor).
- "Simple Application Messaging Protocol," Version 1.3 IVOA Recommendation, 11 April 2012 (M. Taylor, T. Boch, M. Fitzpatrick, A. Allan, J. Fay, L. Paioro, J. Taylor, D. Tody).
- "Simple Line Access Protocol," Version 1.0, IVOA Recommendation, 09 December 2010 (J. Salgado, P. Osuna, M. Guainazzi, I. Barbarisi, M.-L. Dubernet, D. Tody).
- "Simple Spectral Access Protocol," Version 1.1, IVOA Recommendation, 10 February 2012 (D. Tody, M. Dolensky, J. McDowell, F. Bonnarel, T. Budavari, I. Busko, A. Micol, P. Osuna, J. Salgado, P. Skoda, R. Thompson, F. Valdes, and the Data Access Layer working group).
- "Table Access Protocol," Version 1.0, IVOA Recommendation, 27 March 2010 (P. Dowler, G. Rixon, D. Tody).
- "TAPRegExt: a VOResource Schema Extension for Describing TAP Services," Version 1.0, IVOA Recommendation, 27 August 2012 (M. Demleitner, P. Dowler, R. Plante, G. Rixon, M. Taylor).
- "IVOA Spectral Data Model," Version 2.0, IVOA Proposed Recommendation, 09 March 2014 (J. McDowell, D. Tody, T. Budavari, M. Dolensky, I. Kamp, K. McCusker, P. Protopapas, A. Rots, R. Thompson, F. Valdes, P. Skoda, B. Rino, S. Derriere, J. Salgado, O. Laurino, and the IVOA Data Access Layer and Data Model Working Groups).
- "Observation Data Model Core Components and its Implementation in the Table Access Protocol," Version 1.0, IVOA Recommendation, 28 October 2011 (M. Louys, F. Bonnarel, D. Schade, P. Dowler, A. Micol, D. Durand, D. Tody, L. Michel, J. Salgado, I. Chilingarian, B. Rino, J. de Dios Santander, P. Skoda).
- "VOSpace Specification," Version 2.0, IVOA Recommendation, 29 March 2013 (M. Graham, D. Morris, G. Rixon, P. Dowler, A. Schaaff, D. Tody).
- "IVOA Credential Delegation Protocol," Version 1.0, IVOA Recommendation, 18 February 2010 (M. Graham, R. Plante, G. Rixon, G. Taffoni).
- "Web Services Basic Profile," Version 1.0, IVOA Recommendation, 16 December 2010 (A. Schaaff, M. Graham).

- "StandardsRegExt: a VOResource Schema Extension for Describing IVOA Standards," Version 1.0, IVOA Recommendation, 08 May 2012 (P. Harrison, D. Burke, R. Plante, G. Rixon, D. Morris, and the IVOA Registry Working Group).
- "Describing Simple Data Access Services," Version 1.0, IVOA Recommendation, 25 November 2013 (R. Plante, J. Delago, P. Harrison, D. Tody, and the IVOA Registry Working Group).
- "VODataService: a VOResource Schema Extension for Describing Collections and Services," Version 1.1, IVOA Recommendation, 02 December 2010 (R. Plante, A. Stébé, K. Benson, P. Dowler, M. Graham, G. Greene, P. Harrison, G. Lemson, T. Linde, G. Rixon).
- "IVOA Registry Relational Schema," Version 1.0, IVOA Proposed Recommendation, 27 February 2014 (M. Demleitner, P. Harrison, M. Molinaro, G. Greene, T. Dower, M. Perdikeas).
- "IVOA Document Standards," Version 1.2, IVOA Recommendation, 13 April 2010 (R.J. Hanisch, C. Arviset, F. Genova, B. Rino).
- "Sky Event Reporting Metadata," Version 2.0, IVOA Recommendation, 11 July 2011 (R. Seaman, R. Williams, A. Allan, S. Barthelmy, J. Bloom, J. Brewer, R. Denny, M. Fitzpatrick, M. Graham, N. Gray, F. Hessman, S. Marka, A. Rots, T. Vestrand, P. Wozniak).
- "IVOA Support Interfaces," Version 1.0, IVOA Recommendation, 31 May 2011 (Grid and Web Services Working Group, M. Graham, G. Rixon).

Appendix C. International Virtual Observatory Alliance Leadership

Table C.4 lists VAO Team members who served in various leadership positions within the IVOA.

Table C.4: VAO Leadership within the IVOA.

	Position	Individual	Term
Executive committee	Chair	R. Hanisch	2002 June–2003 July
	Deputy Chair	D. De Young	2006 August–2007 August
	Chair	D. De Young	2007 August–2008 October
	Secretary	J. Evans	2013 September–2014 September
Technical Working Group	Chair	R. Williams	2002 June–2006 July
Technical Coordination Group	Chair	R. Williams	2006 July–2008 May
	Deputy Chair	M. Graham	2012 May–2014 September
Inter-operability Conference Program Organizing Committee	Member	R. Hanisch	2003 March–2007 May
	Member	M. Graham	2012 May–2014 September
Standards and Process Subcommittee	Member	R. Hanisch	2007 September–2010 September
Document Coordinator		S. Emery Bunn	2010 July–2014 September
Applications Working Group	Chair	Tom McGlynn	2008 July–2011 July
	Vice Chair	Tom Donaldson	2014 May–2014 September
Data Access Layer Working Group	Chair	Doug Tody	2003 June–2007 May
	Vice Chair	Mike Fitzpatrick	2010 May–2013 May
Data Models Working Group	Chair	Jonathan McDowell	2003 June
	Vice Chair	Omar Laurino	2011 May–2014 May
	Vice Chair	Omar Laurino	2014 May–2014 September
Grid and Web Services Working Group	Chair	Matthew Graham	2006 December–2007 May
	Chair	Matthew Graham	2007 May–2011 May
Registry Working Group	Chair	Ray Plante	2006 September–2009 September
	Chair	Ray Plante	2009 November–2010 November
	Vice Chair	Gretchen Greene	2009 November–2010 November
	Chair	Gretchen Greene	2011 January–2014 May
Standards and Processes Working Group	Chair	Bob Hanisch	2003 June–2006 May

Uniform Content Descriptors Working Group	Chair	Roy Williams	2003 June–2005 January
VO Event Working Group	Chair	Roy Williams	2005 January–2008 January
	Chair	Rob Seaman	2006 December–2008 May
	Chair	Rob Seaman	2008 May–2011 May
	Vice Chair	Roy Williams	2010 October–2011 October
	Chair	Matthew Graham	2011 October–2012 October
Applications Interest Group	Chair	Tom McGlynn	2004 January–2005 July
Data Curation and Preservation Interest Group	Chair	Bob Hanisch	2007 May–2010 May
	Chair	Alberto Accomazzi	2010 May–2014 May
Knowledge Discovery in Databases Interest Group	Chair	George Djorgovski	2012 October–2014 September
Time Domain Interest Group	Chair	Matthew Graham	2012 October–2013 May
	Vice Chair	Mike Fitzpatrick	2013 May–2014 September

The term of the Chair of the Executive Committee was increased to 18 months beginning in 2007 August.

In 2005 July, the Technical Working Group was reformulated as the Technical Coordination Group.

The Chair and Vice Chair of the Data Models Working Group were both granted one year extensions in 2014 May.

The Standards and Processes Working Group was deactivated in 2005 May.

The Uniform Content Descriptors Working Group was renamed to the Semantics Working Group in 2005 October.

The VO Event Working Group was converted to the Time Domain Interest Group in 2012 October.

The Applications Interest Group was converted to the Applications Working Group in 2007 January.

REFERENCES

1. Anderson, K.R., Rosolowsky, E., Dowler, P., 2013. CyberSKA radio imaging metadata and VO compliance engineering. In: Friedel, D.N. (Ed.), Astronomical Data Analysis Software and Systems XXII. In: Astronomical Society of the Pacific Conference Series, vol. 475. p. 231.

2. Arviset, C., Gaudet, S., The IVOA Technical Coordination Group, November 2010. IVOA Architecture, Version 1.0. International Virtual Observatory Alliance. URL: http://www.ivoa.net/documents/Notes/IVOAArchitecture/.

3. Astropy Collaboration, Robitaille, T.P., Tollerud, E.J., Greenfield, P., Droettboom, M., Bray, E., Aldcroft, T., Davis, M., Ginsburg, A., Price-Whelan, A.M., Kerzendorf, W.E., Conley, A., Crighton, N., Barbary, K., Muna, D., Ferguson, H., Grollier, F., Parikh, M.M., Nair, P.H., Unther, H.M., Deil, C., Woillez, J., Conseil, S., Kramer, R., Turner, J.E.H., Singer, L., Fox, R., Weaver, B.A., Zabalza, V., Edwards, Z.I., Azalee Bostroem, K., Burke, D.J., Casey, A.R., Crawford, S.M., Dencheva, N., Ely, J., Jenness, T., Labrie, K., Lim, P.L., Pierfederici, F., Pontzen, A., Ptak, A., Refsdal, B., Servillat, M., Streicher, O., 2013. Astropy: A community Python package for astronomy. A&A 558, A33

4. Berriman, G.B., Good, J.C., Curkendall, D.W., Jacob, J.C., Katz, D.S., Prince, T.A., Williams, R., 2003. Montage: an on-demand image mosaic service for the NVO. In: Payne, H.E., Jedrzejewski, R.I., Hook, R.N. (Eds.), Astronomical Data Analysis Software and Systems XII. In: Astronomical Society of the Pacific Conference Series, vol. 295. p. 343.

5. Boch, T., Donaldson, T., Durand, D., Fernique, P., O'Mullane, W., Reinecke, M., Taylor, M., June 2014. MOC—HEALPix Multi-Order Coverage map, Version 1.0. International Virtual Observatory Alliance, iVOA Recommendation. URL: http://www.ivoa.net/twiki/bin/view/IVOA/IvoaApplications

6. Bonnarel, F., Fernique, P., Bienaymé, O., Egret, D., Genova, F., Louys, M., Ochsenbein, F., Wenger, M., Bartlett, J.G., 2000. The ALADIN interactive sky atlas. A reference tool for identification of astronomical sources. A&AS 143, 33–40.

7. Brunner, R.J., Djorgovski, S.G., Szalay, A.S. (Eds.), 2001. Virtual Observatories of the Future. In: Astronomical Society of the Pacific Conference Series, vol. 225.

8. Busko, I., 2002. Specview: a Java tool for spectral visualization and model fitting of multi-instrument data. In: Starck, J.-L., Murtagh, F.D. (Eds.), Astronomical Data Analysis II. In: Society of Photo-Optical Instrumentation Engineers (SPIE) Conference Series, vol. 4847. pp. 410–418.

9. CASA Consortium, 2011. CASA: Common Astronomy Software Applications. Astrophysics Source Code Library.

10. Davenhall, A.C., Qin, C.L., Noddle, K.T., Walton, N.A., 2004. The astroGrid MySpace system. In: Ochsenbein, F., Allen, M.G., Egret, D. (Eds.), Astronomical Data Analysis Software and Systems (ADASS) XIII. In: Astronomical Society of the Pacific Conference Series, vol. 314. p. 330

11. Demleitner, M., Greene, G., Le Sidaner, P., Plante, R.L., 2014. The virtual observatory registry. Astron. Comput. 7, 101–107. doi:10.1016/j.ascom.2014.07.001.

12. Doe, S., Bonaventura, N., Busko, I., D'Abrusco, R., Cresitello-Dittmar, M., Ebert, R., Evans, J., Laurino, O., McDowell, J., Pevunova, O., Refsdal, B., 2012. Iris: the VAO SED application. In: Ballester, P., Egret, D., Lorente, N.P.F. (Eds.), Astronomical Data Analysis Software and Systems XXI. In: Astronomical Society of the Pacific Conference Series, vol. 461. p. 893.

13. Doe, S., Nguyen, D., Stawarz, C., Siemiginowska, A., Burke, D., Evans, I., Evans, J., McDowell, J., Refsdal, B., Houck, J., Nowak, M., 2006. Sherpa: goals and design for chandra and beyond. In: Gabriel, C., Arviset, C., Ponz, D., Enrique, S. (Eds.), Astronomical Data Analysis Software and Systems XV. In: Astronomical Society of the Pacific Conference Series, vol. 351. p. 77.

14. Dowler, P., Tody, D., Bonnarel, F., Plante, R., July 2014. Simple Image Access Version 2.0. International Virtual Observatory Alliance. URL: http://www.ivoa. net/documents/SIA/.

15. Evans, J.D., Plante, R.L., Boneventura, N., Busko, I., Cresitello-Dittmar, M., D'Abrusco, R., Doe, S., Ebert, R., Laurino, O., Pevunova, O., Refsdal, B., Thomas, B., 2012. Managing distributed software development in the Virtual Astronomical Observatory. In: Society of Photo-Optical Instrumentation Engineers (SPIE) Conference Series, vol. 8449. p. 84490I

16. Freeman, P., Doe, S., Siemiginowska, A., 2001. Sherpa: a mission-independent data analysis application. In: Starck, J.-L., Murtagh, F.D. (Eds.), Astronomical Data Analysis. In: Society of Photo-Optical Instrumentation Engineers (SPIE) Conference Series, vol. 4477. pp. 76–87

17. Geißler, K., Metchev, S., Kirkpatrick, J.D., Berriman, G.B., Looper, D., 2011. A Crossmatch of 2MASS and SDSS. II. Peculiar L Dwarfs, Unresolved Binaries, and the Space Density of T Dwarf Secondaries. Astrophys. J. 732, 56.

18. Graham, M.J., Djorgovski, S.G., Donalek, C., Drake, A.J., Mahabal, A.A., Plante, R.L., Kantor, J., Good, J.C., 2012. Connecting the time domain community with the Virtual Astronomical Observatory. In: Society of Photo-Optical Instrumentation Engineers (SPIE) Conference Series, vol. 8448. p. 84480P.

19. Graham, M.J., Fitzpatrick, M.J., McGlynn, T.A. (Eds.), 2007. The National Virtual Observatory: Tools and Techniques for Astronomical Research. In: Astronomical Society of the Pacific Conference Series, vol. 382

20. Graham, M., Morris, D., Rixon, G., Dowler, P., Schaaff, A., Tody, D., Major, B., 2014. VOSpace Service Specification, Version 2.1. International Virtual Observatory Alliance. URL: http://www.ivoa.net/documents/VOSpace/.

21. Jaeger, S., 2008. The Common Astronomy Software Application (CASA). In: Argyle, R.W., Bunclark, P.S., Lewis, J.R. (Eds.), Astronomical Data Analysis Software and Systems XVII. In: Astronomical Society of the Pacific Conference Series, vol. 394. p. 623.

22. Jeschke, E., Inagaki, T., Kackley, R., 2013. Introducing the ginga FITS viewer and toolkit. In: Friedel, D.N. (Ed.), Astronomical Data Analysis Software and Systems XXII. In: Astronomical Society of the Pacific Conference Series, vol. 475. p. 319.

23. Juric, M., Kantor, J., Axelrod, T.S., Dubois-Felsmann, G.P., Becla, J., Lim, K., LSST Collaboration, , LSST Science Collaborations, , 2013. LSST data products: enabling LSST science. In: American Astronomical Society Meeting Abstracts, vol. 221. p. #247.01.

24. Kinne, R.C., Templeton, M.R., Henden, A.A., Zografou, P., Harbo, P., Evans, J., Rots, A.H., LAZIO, J., 2013. Distributing variable star data to the virtual observatory. In: American Astronomical Society Meeting Abstracts, vol. 221. p. #240.37

25. Laurino, O., Budynkiewicz, J., Busko, I., Cresitello-Dittmar, M., D'Abrusco, R., Doe, S., Evans, J., Pevunova, O., 2014a. Iris: constructing and analyzing spectral energy distributions with the virtual observatory. In: Manset, N., Forshay, P. (Eds.), Astronomical Society of the Pacific Conference Series, 485. p. 19.

26. Laurino, O., Budynkiewicz, J., D'Abrusco, R., Bonaventura, N., Busko, I., CresitelloDittmar, M., Doe, S.M., Ebert, R., Evans, J.D., Norris, P., Pevunova, O., Refsdal, B., Thomas, B., Thompson, R., 2014b. Iris: an extensible application for building and analyzing spectral energy distributions. Astron. Comput

27. Laurino, O., Busko, I., Cresitello-Dittmar, M., D'Abrusco, R., Doe, S., Evans, J., Pevunova, O., 2013. Extending iris: the VAO SED analysis tool. In: Friedel, D.N. (Ed.), Astronomical Data Analysis Software and Systems XXII. In: Astronomical Society of the Pacific Conference Series, vol. 475. p. 295.

28. Laurino, O., D'Abrusco, R., Cresitello-Dittmar, M., Evans, J., McDowell, J., 2012. SedImporter: a tool and an extensible framework for constructing interoperable spectral energy distribution data files. In: Ballester, P., Egret, D., Lorente, N.P.F. (Eds.), Astronomical Data Analysis Software and Systems XXI. In: Astronomical Society of the Pacific Conference Series, vol. 461. p. 391.

29. Mahabal, A., Kembhavi, A., Williams, R., 2011. Astronomy with cutting-edge ICT: from transients in the sky to data over the continents (India-US). In: Proceedings of the Asia-Pacific Advanced Network, vol. 32. p. 143. doi:10.7125/APAN.32.18

30. McGlynn, T.A., 2007. Web-based tools—using the virtual observatory datascope tool. In: Graham, M.J., Fitzpatrick, M.J., McGlynn, T.A. (Eds.),

Astronomical Society of the Pacific Conference Series, vol. 382. p. 51. (Chapter 6).

31. McKee, C.F., Taylor Jr., J.H., Hollenbach, D.J., Boroson, T., Freedman, W., Jewitt, D.C., Kahn, S.M., Moran Jr., J.M., Nelson, J.E., 2001. Astronomy and Astrophysics in the New Millennium. National Academy Press

32. Mishin, D., Medvedev, D., Szalay, A.S., Plante, R., Graham, M., 2014. Data Sharing and Publication Using the SciDrive Service. In: Manset, N., Forshay, P. (Eds.), Astronomical Data Analysis Software and Systems XXIII. In: Astronomical Society of the Pacific Conference Series, vol. 485. p. 465

33. Mockapetris, P., November 1987. RFC 1034: Domain Names—Concepts and Facilities. The Internet Engineering Task Force. URL: http://tools.ietf.org/html/ rfc1034.

34. National Optical Astronomy Observatories, 1999. IRAF: Image Reduction and Analysis Facility. Astrophysics Source Code Library

35. O'Mullane, W., Gray, J., Li, N., Budavári, T., Nieto-Santisteban, M.A., Szalay, A.S., 2004. Batch query system with interactive local storage for SDSS and the VO. In: Ochsenbein, F., Allen, M.G., Egret, D. (Eds.), Astronomical Data Analysis Software and Systems (ADASS) XIII. In: Astronomical Society of the Pacific Conference Series, vol. 314. p. 372

36. Plante, R., Yekkirala, V., Baker, W., 2012. Enabling OpenID Authentication for VOintegrated Portals. In: Ballester, P., Egret, D., Lorente, N.P.F. (Eds.), Astronomical Data Analysis Software and Systems XXI. In: Astronomical Society of the Pacific Conference Series, vol. 461. p. 423

37. Quinn, P.J., Górski, K.M. (Eds.), 2004. Toward an International Virtual Observatory.

38. Salgado, J., Osuna, P., Guainazzi, M., Barbarisi, I., Dubernet, M.-L., Tody, D., December 2010. Simple Line Access Protocol Version 1.0. International Virtual Observatory Alliance. URL: http://www.ivoa.net/documents/SLAP/

39. Seaman, R.L., Vestrand, W.T., Hessman, F.V., 2014. Reengineering observatory operations for the time domain. In: Society of Photo-Optical Instrumentation Engineers (SPIE) Conference Series, vol. 9149. p. 914906

40. Taylor, M.B., 2005. TOPCAT & STIL: Starlink Table/VOTable Processing Software. In: Shopbell, P., Britton, M., Ebert, R. (Eds.), Astronomical Data Analysis Software and Systems XIV. In: Astronomical Society of the Pacific Conference Series, vol. 347. p. 29.

41. Taylor, M., 2011. TOPCAT: Tool for OPerations on Catalogues And Tables. Astrophysics Source Code Library

42. Taylor, M.B., Boch, T., Fay, J., Fitzpatrick, M., Paioro, L., 2012b. SAMP: application messaging for desktop and Web applications. In: Ballester, P., Egret, D., Lorente, N.P.F. (Eds.), Astronomical Data Analysis Software and Systems XXI. In: Astronomical Society of the Pacific Conference Series, vol. 461. p. 279.

43. Taylor, M., Boch, T., Fitzpatrick, M., Allan, A., Fay, J., Paioro, L., Taylor, J., Tody, D., April 2012a. IVOA recommendation: simple application messaging protocol, Version 1.3. International Virtual Observatory Alliance. URL: http://www.ivoa. net/twiki/bin/view/IVOA/IvoaApplications

44. Tody, D., 1986. The IRAF Data Reduction and Analysis System. In: Crawford, D.L. (Ed.), Instrumentation in astronomy VI. In: Society of Photo-Optical Instrumentation Engineers (SPIE) Conference Series, vol. 627. p. 733.

45. Tody, D., 1993. IRAF in the Nineties. In: Hanisch, R.J., Brissenden, R.J.V., Barnes, J. (Eds.), Astronomical Data Analysis Software and Systems II. In: Astronomical Society of the Pacific Conference Series, vol. 52. p. 173

46. Tody, D., Dolensky, M., McDowell, J., Bonnarel, F., Budavari, T., Busko, I., Micol, A., Osuna, P., Salgado, J., Skoda, P., Thompson, R., Valdes, F., Data Access Layer Working Group, February 2012. Simple Spectral Access Protocol Version 1.1. International Virtual Observatory Alliance. URL: http://www.ivoa.net/ documents/SSA/.

47. Tody, D., Plante, R., November 2009. Simple Image Access Specification Version 1.0. International Virtual Observatory Alliance. URL: http://www.ivoa.net/ documents/SIA/20091116/.

48. White, R.R., Allan, A., Barthelmy, S., Bloom, J., Graham, M., Hessman, F.V., Marka, S., Rots, A., Scholberg, K., Seaman, R., Stoughton, C., Vestrand, W.T., Williams, R., Wozniak, P.R., 2006. Astronomical network event and observation notification. Astron. Nachr. 327, 775.

49. Williams, R., Hanisch, R., Szalay, A., Plante, R., February 2008. Simple Cone Search Version 1.03. International Virtual Observatory Alliance. URL: http://www.ivoa. net/documents/latest/ConeSearch.html.

CITATION

R.J. Hanisch, G.B. Berriman, T.J.W. Lazio, S. Emery Bunn, J. Evans, T.A. McGlynn, R. Plante, The Virtual Astronomical Observatory: Re-engineering access to astronomical data, Astronomy and Computing, Volume 11, Part B, June 2015, Pages 190-209, ISSN 2213-1337, http://dx.doi.org/10.1016/j.ascom.2015.03.007.

Index